PENGUIN BOOKS
Gastrophysics

'A fascinating look at the science of food and how our perception is shaped by all our senses, not just taste' *Sunday Times*

'This book really could change the way you eat. There are revelations on every page' *Prospect*

'Highly enlightening. Revealing, very interesting and well worth understanding' *Spectator*

'Packed with new ways of thinking about food' *i*

'A charming guide to empowering ourselves with food science' *Times Literary Supplement*

'Spence offers simple adjustments that can enhance dining pleasure' *Time*

'Compelling and revealing reading' *Universe*

'Fascinating and provocative' *The Times*

'Spence has given us much food for thought' *Independent*

'Truly accessible, entertaining and informative. On every page there are ideas to set you thinking and widen your horizons' Heston Blumenthal, OBE

'This is partly a serious tome and partly an amusing guide for the layperson to a whole new gustatory world. *Gastrophysics* is packed with such tasty factual morsels that could be served up at dinner parties. If Spence can percolate all these factual morsels to the mainstream, the benefits to all of us would be obvious' Nick Curtis, *Daily Telegraph*

'Spence allows people to appreciate the multisensory experience of eating' *New Yorker*

'Fascinating, engaging' *Sunday Times*

'There's much to savour' *Nature*

'Wonderfully curious and thought-provoking, throwing the whole question of why we eat the way we do wide open. Brilliant when demonstrating how much the environment of the table affects our eating – particularly at high-end restaurants' *Guardian*

ABOUT THE AUTHOR

Professor Charles Spence has spent the last two decades researching how people perceive the world around them, earning him the international reputation as the expert in multisensory perception and experience design.

As head of the Crossmodal Research Laboratory at Oxford University, Professor Spence studies how our brains process and integrate the information from each of our senses. He has consulted for many multinational companies, including Unilever, PepsiCo, Diageo, Pernod Ricard, P&G, Nestlé and Twinings, advising on various aspects of multisensory design, packaging, and branding, and has conducted research with a number of world-leading chefs, mixologists and baristas, including Heston Blumenthal and Ferran Adrià.

He has been profiled in publications including the *Guardian*, the *Financial Times* and the *New Yorker*, is a regular on BBC Radio 4's *The Kitchen Cabinet*, and his research has been covered by publications from the *Economist* to the *New Scientist*. In 2008 he was awarded an Ig Nobel Prize for his groundbreaking work on the 'sonic crisp', demonstrating how a louder crunch makes a crisp seem fresher.

His last book, *The Perfect Meal* (written together with Betina Piqueras-Fiszman), won the 2015 Popular Science Prose Award. This is his first trade book.

Gastrophysics

The New Science of Eating

PROFESSOR
CHARLES SPENCE

Foreword by Heston Blumenthal

PENGUIN BOOKS

PENGUIN BOOKS

UK | USA | Canada | Ireland | Australia
India | New Zealand | South Africa

Penguin Books is part of the Penguin Random House group of companies
whose addresses can be found at global.penguinrandomhouse.com.

First published by Viking 2017
Published with an abridged Notes section in Penguin Books 2018
001

Copyright © Charles Spence, 2017
Introduction copyright © Heston Blumenthal, 2017

The moral right of the copyright holders has been asserted

Typeset by Jouve (UK), Milton Keynes
Printed in Great Britain by Clays Ltd, St Ives plc

A CIP catalogue record for this book is available from the British Library

ISBN: 978–0–241–97774–3

www.greenpenguin.co.uk

To Norah Spence, who knew implicitly the value of a good education without ever having had the opportunity to have one.

And Barbara Spence, who had to read more about the legendary F. T. than any loving wife should ever have to.

Contents

Acknowledgements

I would never have ended up in the world of gastrophysics if it hadn't been for the enduring support and mentorship of Prof. Francis McGlone then at Unilever Research, for which I will always remain grateful. As will become clear from the main text, though, it was really the introduction to Heston Blumenthal by Tony Blake of Firmenich that led to my growing interest in gastronomy, rather than food science! In recent years, I owe an especial debt of gratitude to Rupert Ponsonby (R&R), Christophe Cauvy (then of JWT), and Steve Keller (iV Audio Branding) for having believed in the multisensory approach to gastrophysics and all things fun. To Prof. Barry Smith, for helping make the Baz 'n' Chaz wine roadshow so enjoyable. Long may it continue! It has, though, really been the enthusiastic support and collaboration of the next generation of young chefs, including Jozef Youssef, of Kitchen Theory, and Charles Michel, Crossmodalist extraordinaire, that has made the latest gastrophysics research such fun to do. You will read about a number of their dishes and designs in the pages that follow.

I would also like to thank the many chefs and culinary schools for their support, and opening up their kitchens and restaurants to the 'Mad Professor': I have been lucky enough to conduct gastrophysics research over the last fifteen years together with a number of world-leading chefs including Heston Blumenthal and all the team at The Fat Duck Research Kitchen and restaurant; Chef Sriram Aylur, Quilon, London; Chef Jesse Dunford Woods, Parlour, London; Ben Reade, Nordic Food Lab; Dominique Persoone, The Chocolate Line; Chef Albert Landgraf from Epice, São Paolo; Chef Xavier Gamez, of Xavier260, Porto Allegre, Brazil; Chef Andoni and Dani Lasa from Mugaritz, San Sebastián; Chef Joel Braham, of The Good Egg, London; Chef Debs Paquette, of Etch, Nashville; and not forgetting Chef Paul Fraemohs, of Somerville College, Oxford. I have also been lucky enough to conduct research together with Ferran Adrià's Alicía Foundation in Spain, The Paul Bocuse Cookery School, Lyon, France, and Westminster Kingsway College, London. I would also like to thank Jelly & Gin, Blanch & Shock, Caroline Hobkinson, Sam Bompas, and all the students, past and present, who have done most of the research here at the Crossmodal Research Laboratory.

Finally, I would like to thank Tony Conigliaro from 69 Colbrooke Row, London, Ryan Chetiyawardana, aka Mr Lyan, Neil Perry (of Rockpool, Sydney), and Maxwell Colonna-Dashwood, of Colonna & Small's, Bath. All masters of their art. And, finally, Fergus Henderson, for the memorable evening onstage at the Cheltenham Science Festival back in 2007 (along with a bucket of tripe oh so gallantly displayed by my vegan then graduate student, Maya Shankar).

Foreword

There was a time when – apart from the late, great Nicholas Kurti – scientists didn't consider the science of food a serious or worthwhile subject for study. I'd talk with them, offering up theories based on what I'd observed and carefully tested in The Fat Duck kitchen, and get an indulgent smile that seemed to say, 'You stick to cooking and let us get on with the rest.' Admittedly, chefs were no better, insisting that cooking had little to do with science, as though the eggs they were busy scrambling weren't in fact undergoing the technical process of coagulation.

Charles, though, wasn't like this. One of his strengths is that he has a curiosity that crosses disciplines and, for all his scientific rigour, isn't confined to a narrow academic viewpoint. Upon meeting him, I discovered that many of the ideas I was exploring in my kitchen, he was also exploring in his lab. And so, as you'll see in this book, he and I began doing research together on how we react to the food we see, hear, smell, touch, and put in our mouths. We eat with our eyes, ears, nose, memory, imagination and our gut. Every human being has a relationship with food, some of it positive, some of it negative, but ultimately it's all about emotion and feeling.

To me, this is at the very heart of how we respond to food: much more than the tongue (which detects at least five tastes); more even than the nose (which detects countless aromas), it's the conversation between our brain and our gut, mediated by our heart, that tells us whether we like a food or not. It's the brain that governs our emotional response.

It's a hugely rewarding subject (and an essential one for us, as humans, to understand), but it's undoubtedly a complex one, too. Charles is the perfect guide to introduce us to this world and to investigate with us – in a truly accessible, entertaining and informative way – how it works. On every page there are ideas to set you thinking and widen your horizons, from the notion that we all of us live in separate and completely different taste worlds, to questions like, 'Is cutlery the best way to move the food from plate to mouth?'

What I take away from *Gastrophysics* is that, as Charles says, in the mouth very little is as it seems. The pleasure we get from food depends, far more than we could possibly imagine, on our subjectivity – on our memories, associations and emotions. It's a fascinating topic into which you can take your first steps through the door by reading *Gastrophysics*.

Heston Blumenthal

Amuse Bouche

'Open wide!' she said, in her most seductive French accent, and so I did. And in it went. In that one moment, in that one movement, and in that one mouthful, I was taken back to the haziest memories of being spoon-fed as a baby (or at least my imagining of what that must have been like). That dish, or rather the way in which it was served, also foreshadowed what my last meals may well be like as the darkness draws in. So, if you want just one example to illustrate how food is so much more than merely a matter of nutrition, then that was it – that mouthful of lime *gelée* at The Fat Duck restaurant in Bray, many, many years ago. It was an incredibly powerful experience, shocking, disturbing even. But why? Well, I guess in part because no one had fed me that way, at least not in the last forty-five years or so.* Yet there I was, at what was soon to become the world's top restaurant, being spoon-fed my three-Michelin-starred dinner. Well, one course of it, at least. Just enough to make the point that dining is about much more than merely what we eat.

The pleasures of the table reside in the mind, not in the mouth. Get that straight and it soon becomes clear why cooking, no matter how exquisitely executed, can only take you so far. One needs to understand the role of 'the everything else' in order to determine what really makes food and drink so enjoyable, stimulating and, most importantly, memorable. Even something as simple as biting into a

---

* The only thing missing was for the waitress to have me sit on her lap before serving me! I doubt that Heston and the gang would dare to repeat this little interlude at The Duck today. It is just a little too provocative, a little too 'out there' for those gastrotourists who can afford the £295 price of admission, now that the restaurant has firmly established itself as one of the grand temples of modernist cuisine. But others have since taken up the baton from Heston and his ilk, such as Dabiz Muñoz, the 'bad boy' of modernist cuisine, at DiverXo, in Madrid.

fresh ripe peach turns out, on closer inspection, to be an incredibly complex multisensory experience. Just think about it for a moment: your brain has to bind together the aromatic smell, the taste, the texture, the colour, the sound as your teeth bite through the juicy flesh, not to mention the furry feeling of the peach fuzz in your hand and mouth. All of these sensory cues, together with our memories, contribute much more than you would believe to the flavour itself. And it all comes together in your brain.

It is the growing awareness that tasting is fundamentally a cerebral activity that is leading some of the world's top chefs to take a fresh look at the experiences that they deliver to their diners. Just take Denis Martin's modernist restaurant in Switzerland (see Figure 0.1). The chef realized that some of his guests were not enjoying the food as much as he thought they should, given how much effort he was putting into preparing the dishes. Too often his diners were stiff and buttoned-up – 'Suits on account', as he put it. How could anyone who walked in the door sporting such a sour expression be expected to enjoy his food? The solution was brilliantly simple, and involved putting a cow on each and every table.

Nothing happens at the start of service until one of the diners, curious as to whether what they see before them on the table is a Swiss take on a salt shaker or pepper grinder, picks up their cow. When they tilt it to look underneath, it lets out a mournful moo. Diners often laugh in surprise. Then, within a few moments, the dining room erupts into a chorus of mooing cows, and the restaurant is full of chortling diners. The mood has been lifted and that is when the first course comes out from the kitchen.* This wonderfully intuitive *mental* palate cleanser is far more effective than any acidic sorbet – the traditional means of cleansing the palate – at enhancing the diners' enjoyment of the food to come. After all, our mood is one of the most important factors influencing our dining experience, so best try to optimize it.

It turns out that modernist chefs are especially interested in the

---

* Notice also how this helps bring the diners in the restaurant together in a shared sonic experience (see 'Social Dining').

Figure 0.1. The only item of tableware to greet the expectant diner at Denis Martin's two-Michelin-starred restaurant in Vevey, Switzerland. But what exactly are you looking at, and why has the chef placed one on each and every table?

new sciences of eating (what I will here call gastrophysics), given their habit of recombining ingredients in new and unusual ways, not to mention their desire to play with diners' expectations. How exactly they are using this emerging knowledge to enhance the experience of eating constitutes the subject matter of this book. Many of the food and drinks companies are also becoming increasingly curious about the science of multisensory flavour perception. The aspirations of the latter, though, tend to be somewhat different from those of the chefs. Their hope is that the new gastrophysics insights may help them to use the so-called 'tricks of the mind' in order to reduce some of the unhealthy ingredients in their branded food products without having to compromise on taste.

Gastrophysics: The new sciences of eating

Many factors influence our experience of food and drink, whether we are eating something as simple as a luscious ripe peach or a fancy dish at one of the world's top restaurants. However, none of the existing approaches provides a complete answer as to why food tastes the way it does and why we crave some dishes but not others. After all, the focus of modernist cuisine is primarily on food and its preparation – often described as the new science of the kitchen. Sensory science, meanwhile, tells us about people's perception of the sensory attributes of what they eat and drink in the lab, how sweet the taste, how intense the flavour, how much they like the dish. And then there is neuro-gastronomy – basically, the study of how the brain processes sensory information relating to flavour. This new discipline helps shed light on the brain networks that are involved when people taste liquidized food pumped into their mouth via a tube while lying flat on their back with their head clamped in a brain scanner. Do I have any volunteers? Interestingly, you now find mention of the diner's brain on the menu at top restaurants like Mugaritz in San Sebastián in Spain, and at The Fat Duck restaurant in Bray. In fact, many of the science-inspired trends one now sees coursing through restaurants across the globe can be traced back to Bray, where Heston Blumenthal and his research team, together with their many collaborators, have been pushing the boundaries of what dining could be for more than two decades now.

However, neither modernist cooking, nor sensory science nor even neurogastronomy offers a satisfactory explanation as to why our food experiences, be they special occasion or mundane everyday meal, appear to us as they do. What is needed is a new approach to measuring and understanding those factors that influence the responses of *real* people to *real* food and drink products, ideally under as *naturalistic* conditions as possible. Gastrophysics builds on the strengths of a number of disciplines, including experimental psychology, cognitive neuroscience, sensory science, neurogastronomy, marketing, design and behavioural economics, each subject contributing a part of the story with specific techniques designed to answer particular questions.

As an experimental psychologist, I have always been interested in the senses, and in applying the latest insights from cognitive neuroscience to help improve our everyday experience. While I started out investigating sight and sound, over the years I have been slowly adding more senses to my research. Eventually this led me to the study of flavour, which is, after all, one of the most multisensory of our experiences. Given that my parents never went to school (they were constantly moving around the country, as they grew up on the fairground), I have always had a clear sense that research findings need ultimately to have real-world application. In 1997, I started my lab, the Crossmodal Research Laboratory, which is nowadays funded largely by the food and beverage industry. There are psychologists, obviously, but also marketers, the occasional product designer, musicians, and we even have a Chef in Residence. (Guess who has the tastiest lab parties in Oxford!) I have also been lucky enough to work with leading chefs, mixologists and baristas, and for my tastes the most exciting gastrophysics research lies at the intersection of these three areas – the food and beverage industry, the culinary experience designers and the gastrophysicists. I believe that gastrophysics research will come to play a dominant role in understanding and improving all of our food and drink experiences in the years to come.

What is 'gastrophysics'?

Gastrophysics can be defined as the scientific study of those factors that influence our multisensory experience while tasting food and drink. The term itself comes from the merging of 'gastronomy' and 'psychophysics': gastronomy here emphasizes the fine culinary experiences that are the source of inspiration for much of the research in this area, while psychophysics references the scientific study of perception. Psychophysicists like to treat the human observer much like a machine. By systematically observing how people respond to carefully calibrated sets of sensory inputs, the psychophysicist hopes to measure what their participants (or observers) perceive, and then to figure out what really matters in terms of influencing people's behaviour.

Generally speaking, gastrophysicists aren't interested in simply asking people what they think. Better to focus on what people actually do, and how they respond to specific targeted questions and ratings scales, such as: How sweet is the dessert (give me a number from 1 to 7)? How much did you enjoy the food? How much would you pay for a dish like the one you have just eaten? They tend to be sceptical of much of what people say in unconstrained free report, given the many examples where people have been documented to say one thing but to do another (see 'The Atmospheric Meal' chapter for some great examples of this).

Importantly, the findings of the gastrophysics research do not apply only to high-end food and beverage offerings. If they did, they would still be interesting, certainly, but perhaps just not all that relevant in the grand scheme of things. How often do most of us get to dine at a Michelin-starred restaurant anyway? But many of the modernist chefs are incredibly creative. What is more, they have the authority and capacity to instigate change. If they are intrigued by the latest findings from the gastrophysics lab, they can probably figure out a way of putting a dish inspired by the new science on the menu next week. The large food and beverage companies, by contrast, often find it harder to engage in rapid, not to mention radical, innovation, much though they would like to. In the food industry, everything just tends to happen at a much slower pace!

In the best-case scenario, some of the most inventive ideas first trialled in the modernist restaurant provide genuine insights that can subsequently be used to enhance the experience of whatever we might be eating or drinking, whether we are on an aeroplane or in hospital, at home or in a chain restaurant. The multisensory dishes and experiences first dreamed up in some of these top dining venues provide the proof-of-principle support that gives others the confidence to innovate for the mainstream. So when the collaboration works well, it can lead to emerging gastrophysics insights being turned into amazing food and drink experiences that people really want to talk about and share. Get it right and it can result in dishes that are more sensational, more memorable and possibly healthier than anything that has gone before.

For example, just take the research that we conducted together with Unilever fifteen years ago. We demonstrated that if we boosted the sound of the crunch when people bit into a potato crisp we could enhance

Figure o.2. Chef Heston Blumenthal gets to grip with the 'sonic chip' in the golden booth at the Crossmodal Research Laboratory in Oxford, *c*.2004.

their perception of its crunchiness and freshness. Research, I am proud to say, that led to our being awarded the Ig Nobel Prize for Nutrition. This isn't the same as the Nobel Prize, but a rather more tongue-in-cheek award for science that first makes you laugh, and then makes you think. It was around this time that the chef Heston Blumenthal started coming up to the lab in Oxford, having been introduced by Anthony Blake of the Swiss flavour house Firmenich. As soon as we stuck the headphones on Heston and locked him away in the booth, he got it (see Figure o.2)!

In fact, when interviewed on a BBC Radio 4 show at the time the chef stated: 'I would consider sound as an ingredient available to the chef.' This realization, in turn, provided the original impetus that led to the 'Sound of the Sea' seafood dish, at The Fat Duck, which became the signature dish at one of the world's top restaurants. Other restaurants and brands then started working on adding a sonic element to their dishes, often facilitated by technology.

Subsequently, we worked together with The Fat Duck Research Kitchens on sonic seasoning – basically, a way of systematically modifying the taste of foods by playing specific kinds of sound or music. These insights eventually made their way on to the menu at The House of Wolf restaurant in north London, courtesy of culinary artist Caroline Hobkinson. Culinary artists are more artist than chef, but use food and food installations to express themselves and their ideas. And it was on the basis of such research that British Airways launched their 'Sound Bite' menu in 2014, providing the option of sonic seasoning for their long-haul passengers. More recently still, a number of health authorities have started to research whether they can generate 'sweet-sounding' playlists to help, for example, those diabetic patients who need to control their sugar intake – the idea being that if you can 'trick' the brain into thinking that the food is sweeter than it actually is, you get better-tasting food without the harmful side effects of consuming too much sugar. From the gastrophysics lab to the modernist restaurant, and on to the mainstream (though I worry that the follow-up studies have yet to be done to check just how long-lasting the effects of music and soundscapes are). And it may be that the direction of travel is reversed, with some of what is already going on in the top restaurants providing the impetus for the basic research back in the lab.

What's the difference between 'crossmodal' and 'multisensory'?

Many of the insights of gastrophysics are built on the latest findings coming out of crossmodal and multisensory science. Now, these complex-sounding terms describe the fact that there is much more interplay between our senses than previously thought. While scientists used to think that what we see goes to the visual brain, what we hear to the auditory brain, and so on, it turns out that there are far more connections between the senses than we ever realized. So changing what a person sees can radically alter what they hear, changing what they hear may influence what they feel, and altering what they feel can modify what they taste. Hence the term 'crossmodal', implying that what is going on in one sense influences what we experience in the others (as, for example, when someone puts on some red lighting and suddenly the wine in your black glass tastes sweeter and fruitier).

The term 'multisensory', by contrast, is more often used to explain what happens when, say, I change the sound of the crunch you hear as you bite into a crisp. In the latter case, what you hear and feel are integrated in the brain into a multisensory perception of freshness and crispness, with both senses intrinsic to your experience of one and the same food item. Don't worry if the distinction sounds like a subtle one – it is. Nevertheless, is just the this sort of thing that gets my academic colleagues fired up.

I would certainly like to take issue with the conceit of one recent BBC TV show (*Chef vs Science: The Ultimate Kitchen Challenge*) in the UK, in which chef was set against scientist. Ridiculous, if you ask me. For no matter whether the competition is between Pierre Gagnaire and Hervé This (one of the godfathers of molecular gastronomy), or Michelin-starred *MasterChef* regular Marcus Wareing versus materials scientist Mark Miodownik, the answer isn't really in any doubt – stick with the chef. What is much more interesting, at least to me, is how much of a lift the chef, molecular mixologist or barista can get by working together with the gastrophysicist. In the chapters that follow, I hope to convince you that, more often than not, the combination will win out. Not only that but the fruits of this collaboration are starting to percolate down to influence our food and drink experiences no matter where we eat and regardless of what we choose to consume.

Not everyone is happy about what they see happening in the world of gastronomy, though. *MasterChef* judge William Sitwell, for instance, promised to destroy any square plates you brought to him.[1] He absolutely hates the new fashion in plating. Don't get me wrong, I understand where he is coming from. There are undoubtedly some practitioners out there who have definitely lost the plot. You know what I mean – when the dish you ordered arrives at the table served in a mini frying pan, atop a plank suspended between a couple of bricks. But let's be clear about this: the mere fact that some people take things too far does not invalidate the more general claim that our perception of, and our behaviour around, food is influenced by the way in which it is plated and what it comes served on. What is particularly exciting to me is that one can take some of the latest trends in plating from the high end of gastronomy and translate them into actionable insights that hold the promise of enhancing the food service offering in, for instance, a hospital setting.

Is cutlery the best way to move the food from plate to mouth?

How much do you really like the idea of sticking something into your mouth that has been inserted into who knows how many other mouths beforehand? Think about it carefully – is a cold, smooth

Figure 0.3. Will the tableware of the future look like this? A selection of utensils created by silversmith Andreas Fabian in collaboration with Franco-Colombian chef Charles Michel, as displayed at the 'Cravings' exhibition at London's Science Museum.

stainless steel knife, fork and/or spoon really the best way to transfer food from table to mouth? Why not eat with your fingers instead? Is it mere coincidence that this is how one of the world's most popular foods – the burger – is typically eaten? Given what we now know about the workings of the human mouth and the integration of the senses that give rise to multisensory flavour perception, shouldn't we all think about designing things a little differently, moving forward? Why not give spoons a texture to caress the tongue and lips? After all, the latter are amongst the body's most sensitive skin sites (at least of those that are accessible while seated at the dining table).

Why not cover the handles of one's cutlery with fur, much like the Italian Futurists might have done at their tactile dinner parties in the 1930s? We have tried both here in Oxford (see Figure 0.3). There is inertia to change, certainly. But since we have (mostly) accepted such radical innovations to our plateware in recent years, why not do the same with our cutlery? This question holds true no matter whether your implements of choice happen to be Western cutlery or chopsticks. Excitingly, gastrophysicists are now working with cutlery makers, industrial designers and chefs in order to deliver a better offering to the table.

I am convinced that change really is possible in the world of food and drink, and that progress will come at the interface between modernist cuisine, art and design, technology and gastrophysics. Thereafter, the best ideas will be disseminated out to the mainstream by the food and beverage industry. And by chefs . . . and eventually by you.

Testing intuitions

What the gastrophysics research often does, then, is assess people's intuitions. Typically, the results provide empirical support concerning the relative importance of various different factors that people already suspected were somehow relevant. However, on occasion, the research can turn up a surprise result, one that may, for instance, show that some age-old kitchen folklore is just plain wrong. Let me give you a concrete example to illustrate the point: many chefs are taught in cookery school to place an odd rather than even number of elements on the plate (i.e., serve three scallops or five, rather than four). However, when we tested this practice by showing several thousand people pairs of plates of food and asking them which they preferred (see Figure 0.4 for an example), it really didn't matter. Instead, people's choices correlated to the total amount of food that was on the plate. The more food, the better! Of course, even when the gastrophysics research simply backs up people's intuitions, it can nevertheless help put a monetary value on something, which often aids in decision-making (i.e., is the extra effort/cost of doing things a particular way really worth the effort?).

In the remainder of this introduction, I want to focus on some of the questions that gastrophysicists are currently thinking about, and

Figure o.4. Which plate of seared scallops do you prefer? The latest research shows that we care more about how much food there is than whether there happens to be an odd or an even number of elements on the plate.

bringing to the public's attention. These are some of the key themes that will be discussed in the chapters that follow.

Just how much influence does the atmosphere really have?

Now, whenever we eat, be it in a dine-in-the-dark or Michelin-starred restaurant, the atmosphere, the sights, the sounds, the smell, even the feel of the chair we happen to be sitting on (not to mention the size and shape of the table itself), all influence our perception and/or our behaviour, however subtly. From what we choose to order in the first place to what we think about the taste of the food when it comes, the speed at which we eat and the duration of our stay, the atmosphere affects everything. People will tell you that they were always going to choose what they ordered and to eat and drink as much as they actually did. However, the emerging gastrophysics research shows that this is simply not the case.

In our research with the food and beverage industry, we have been quantifying just how much of an impact the atmosphere really has on people's ratings of taste, flavour and preference. We found, for example, that people's ratings of one and the same drink may vary by 20% or more as a function of the sensory backdrop where it is served. No wonder, then, that – as we will see later – top chefs and restaurateurs are increasingly recognizing the importance of such environmental effects. In some cases, they have sought to match the atmosphere to the food they serve, the image that they wish to create or the emotion that they wish to provoke. In the 'Airline Food' chapter, for instance, we will take a look at how our growing understanding of the impact of the atmosphere on multisensory taste perception is now enabling some of the world's most forward-thinking airlines to improve their food offering at 35,000 feet.

Have you heard of off-the-plate dining?

One of the trends that has been sweeping high-end modernist dining in recent years is the growing focus on off-the-plate dining (see 'The Experiential Meal'). This term is used to describe the more theatrical, magical, emotional, storytelling elements that one increasingly finds in contemporary haute cuisine. Nowadays, it all seems to be about delivering meaningful, memorable and stimulating multisensory experiences (or journeys); selling 'the experience', the *total* product and not just the *tangible* product in Philip Kotler's marketing terminology. Better still if those experiences also happen to be shareable (e.g., for the millennials on their social media).

And while the tops chefs fight over who should get the credit for first coming up with the idea of multisensory experiential theatrical dining, the irony is that the Italian Futurists were already matching meals to sounds eighty years ago, not to mention adding scents and textures to their dinners, and they were amongst the first to experiment with miscolouring the foods that they served. We'll take a closer look at whether modernist cuisine really was invented back in the 1930s in the final chapter ('Back to the Futurists').

Doesn't good food speak for itself?

Some commentators, including a few Michelin-starred chefs, dismiss gastrophysics as nothing more than 'sensory trickery'. 'Good food,' you hear them proclaim oh-so-earnestly, 'should speak for itself.' To them, a great meal is all about the local sourcing, the seasonality of the ingredients, the detail and technique in the preparation, and the beautiful cooking. Don't mess with the food; keep it simple, keep it slow, even. This was certainly the line I heard from Michael Caines MBE, then the Michelin-starred chef at Gidleigh Park in Devon, when I met him in 2015.[2] He'd have you believe that none of this other stuff matters, that the world would – heaven forbid – perhaps be a better place without gastrophysics.

According to the likes of Caines,★ the honest chef lets their dishes do the talking. They don't need to worry about the weight of the cutlery to make *their* food taste great. And yet I don't need to go to Gidleigh Park to know that the cutlery will be heavy. There is just no way that any self-respecting chef would ever serve their food with a plastic or aluminium knife and fork. It would spoil the experience! Tell me, am I wrong? And, hold on a minute, let's take a look at the decor and context. Gidleigh Park just so happens to be a beautiful manor house set in the heart of the Devonshire countryside. I am sure that you don't need a gastrophysicist to tell you that the chef's dishes are going to taste better there than if exactly the same food were to be served in a noisy aeroplane cabin or in a hospital canteen. In other words, you cannot avoid 'the everything else', however much you might wish to.

My point, then, is that wherever food and drink is served, sold or consumed there is always a multisensory atmosphere. And that environment impacts both what we think about what we are tasting and,

★ Caines is very much at the slow food rather than the molecular gastronomy end of the spectrum; and, truth be told, there are far worse places to practise slow food than rural Devon. My problem with the slow food movement, though, is that most of those who advocate it tend to have the luxury of living close to the verdant countryside.

more importantly, how much we enjoy the experience. Ultimately, there is just no such thing as a neutral context or backdrop. It is time to accept the growing body of gastrophysics evidence demonstrating that the environment, not to mention the plateware, dish-naming, cutlery and so on, *all* exert an influence over the tasting experience. Once you have got that straight, then surely it makes sense to try and optimize 'the everything else', along with whatever you happen to be serving on the plate. And this holds true no matter what one is trying to achieve, be it a more memorable, a more stimulating or a healthier meal. Or, I suppose, you can simply stick your head in the sand and pretend that none of this other stuff really matters. To me, the choice is clear. (And my advice for those who choose to ignore all that the emerging science of gastrophysics has to offer is to simply make sure that you are serving your food in a fancy venue with your diners holding heavy cutlery!)

So, without further ado, having polished off the amuse bouche (not to mention the naysayers), let's move on to the first course!

1. Taste

Can you list all of the basic tastes? There is sweet, sour, salty and bitter, for sure. But anything else? Nowadays, most researchers would include umami as the fifth taste. Umami, meaning 'delicious taste', was first discovered back in 1908 by Japanese researcher Kikunae Ikeda. This taste is imparted by glutamic acid, an amino acid, and is most commonly associated with monosodium glutamate, itself a derivative of glutamic acid. Some would be tempted to throw metallic, fatty acid, kokumi and as many as fifteen other basic tastes into the mix as well – though even I haven't heard of most of them. And some researchers query whether there are even any 'basic' tastes at all!

The mistake that many people make, though, when talking about food and drink is to mention things like fruity, meaty, herbal, citrusy, burnt, smoky and even earthy as tastes. But these are not tastes. Strictly speaking, they are flavours. Don't worry, most people are unaware of this distinction. But how do you tell the difference? Well, hold your nose closed – and what is left is taste (at least assuming that you are not tasting something with a trigeminal hit, like chilli or menthol, which activate the trigeminal nerve). So if we struggle to get the basics straight, what hope is there when it comes to some of the more complex interactions taking place between the senses? Taste would be simple, if it weren't so complicated!

Do you mean taste or do you mean flavour
(and does it really matter)?

Most of what people call taste is actually flavour, and many of the things that they describe as flavours turn out, on closer inspection, to be tastes. Some languages manage to sidestep the issue by using the same word for both taste and flavour. In fact, in English, what we really need is to create a new word – and that neologism is 'flave'. 'I love the *flave* of that Roquefort' would do the trick. Let's see whether it catches on. There are also challenges here from those stimuli that lie on the periphery. Just take menthol, the minty note you get when chewing gum: is it a taste, a smell or a flavour? Well, all three, in fact; and it also gives rise to a distinctive mouth-cooling sensation. The metallic sensation we get when we taste blood also has the researchers scratching their heads in terms of whether it should be classified as a basic taste, an aroma, a flavour or some combination of the above.

Most people have heard of the 'tongue map'. In fact, pretty much every textbook on the senses published over the last seventy-five years or so includes mention of it. The basic idea is that we all taste sweet at the front of the tongue, bitter only at the back, sour at the side, etc. However, the textbooks are wrong: your tongue does not work like that! This widespread misconception resulted from a mis-translation of the findings of an early German PhD thesis that appeared in a popular North American psychology textbook written by Edwin Boring in 1942.[1] So now we have got that cleared up, let me ask, do you actually have any idea how the receptors are laid out on your tongue? No, I didn't think so. Something so fundamental, so important to our survival, and yet none of us really has a clue about how it all works. Shocking, no?

The taste receptors are not evenly distributed, but neither are they perfectly segmented as the oft-cited tongue map would have us believe. The answer, as is so often the case, lies somewhere in between. Each taste bud is responsive to all five of the basic tastes. But these taste buds are primarily found on the front part of the tongue, on the sides towards the rear of the tongue and on the back of the tongue. There are no taste buds in the middle of the tongue. Interestingly, though, many people (including chefs) tend to say that they experience sweetness more towards the tip of the tongue, they feel the sourness on the sides of the tongue and bitterness/astringency often seems more noticeable towards the back of the tongue. And for me, a pure umami solution has a mouth-filling quality to it that none of the other tastes can quite match.

The *real* question, though, is how have so many people been so wrong for so long? Part of the reason may be due to the general neglect of the 'lower' senses by research scientists. Another factor probably relates to the 'tricks' that our mind plays on us when constructing flavour percepts, things like 'oral referral' and 'smelled sweetness' (about which more later). As we will see time and again throughout this and the following chapter, in the mouth, very little is as it seems.

Managing expectations

Why, you might well ask, does a cook – be they a modernist chef working in a high-end Michelin-starred restaurant or you slaving away in the kitchen preparing for your next dinner party – need to know about what is going on in the mind of the diners they serve? Why not simply rely on the skills that are taught in the culinary schools or picked up from watching those endless cookery shows on TV? Why not focus on the seasonality, the sourcing, the preparation, and possibly also the presentation of the ingredients on the plate? That is all you need, isn't it? As a gastrophysicist, I know just how important it is to get inside the mind of the diner in order to understand and manage their expectations about food. It is only by combining the best food with the right expectations that any of us can hope to deliver truly great tasting experiences.

It really excites me to see a growing number of young chefs starting to think more carefully about feeding their diners' *minds* and not just their *mouths*. I'm sure this is largely down to the influential role of star chefs like Ferran Adrià and Heston Blumenthal, both of whom I have been lucky enough to work with. Where they lead, others surely follow. But that still doesn't answer the more fundamental question of what got the top chefs interested in the minds of their diners in the first place. After all, this certainly isn't something that they teach you at cookery school.

In Heston's case, it all started with an ice cream. In the late 90s, Heston created a crab ice cream to accompany a crab risotto. The top chef liked the taste and, after a little tinkering, believed it to be perfectly seasoned. But what would the diners say? (Typically, any new dish is trialled in the research kitchen across the road from the restaurant. Then, once it has met with Heston's approval – a slow and exacting process – the next step is to try the new dish out on a few of the regulars and see how they like it. Only if a dish passes all of these hurdles will it stand a chance of making its way on to the restaurant's tasting menu.)

Imagine the scene: just like in one of the chef's TV shows, you can almost see Heston looking on expectantly from the kitchens, waiting for the diners' approval as his latest culinary creation is brought out to the guinea pigs sitting at the tables. Surely the diners will think it tastes great, given who made it. But, in this case at least, the response was not what was expected. 'Urrrggghhh! That's disgusting. It's way too salty.' Well, maybe I exaggerate a little – but trust me, the response wasn't good.

What had gone wrong? How could one of the world's top chefs consider a dish to taste just right only to have some of his regular guests find it far too salty? The answer, I think, tells us a lot about the importance of expectations in our experiences of food and drink. In other words, it is as much a matter of what is in the *mind* of the person doing the tasting as what is in their *mouth* or on the plate. When the diners saw that pinkish-red ice cream (this was also evaluated in the lab with a smoked salmon ice cream), their minds immediately made a prediction about what they had been given to eat. Tell me, what would *you* expect to taste were such a dish to be placed before you?

For most Westerners, pinkish-red in what looks like a frozen dessert is associated with a sweet fruity ice cream, probably strawberry flavour. 'Sweet, fruity, I like it, but it isn't so good for me' – all that goes through a diner's mind in the blink of the eye. After all, one of our brain's primary jobs is to try to figure out which foods are nutritious and worth paying attention to (and perhaps climbing a tree for), and which are potentially poisonous and hence best avoided. However, on the rare occasions when our predictions turn out to be wrong, the surprise, or 'disconfirmation of expectation', that follows can come as quite a shock. It can, in fact, be rather unpleasant. The diners in Heston's restaurant presumably thought that they were going to taste something *sweet*, but what was brought out from the kitchen was actually a *savoury* frozen ice. In other words, they were expecting strawberry and got frozen crab bisque instead! The savoury ice may have been popular in England a century ago, but it has very much fallen out of favour these days.

In a great series of gastrophysics experiments, Martin Yeomans and his team at the University of Sussex, together with Heston, showed that it was possible to radically influence people's perception and liking of the frozen pink treat simply by changing the name of the dish. All it took to modify the participants' expectations in the lab setting was to tell them that this was a savoury ice, or else give the dish the mysterious title 'Food 386'. The expectations that go with the name or description of the dish led people to enjoy the ice cream significantly more than those who had not been told anything about the dish before tasting it. Crucially, they no longer found it too salty either.

Research suggests that our first exposure to a flavour affects those that come thereafter, even once we know exactly what it is that we are tasting. And though the effects may not always be as dramatic as in the case of Heston's pink savoury ice cream, we have probably all had our own experience of this. I still remember, on my first trip to Japan, fifteen years ago, buying a pale-green ice cream from a street vendor. It was a hot spring day and everyone seemed to have one of these refreshing-looking ices in their hands. I had absolutely no doubt that it was mint-flavoured, just as it would be back in the UK. But I recoiled in shock on tasting what turned out to be something most unexpected; it was, in fact, green-tea-flavoured ice cream. Delicious in its way, but I must confess that I have somehow never been able to quite get over that initial surprise whenever I am served a bowl in Japan.

Whatever the name and/or description of a dish, and no matter what it looks like, these cues are always there, helping to set our expectations. And those expectations influence our judgements and perception, however subtly. Even when cooking at home, how those you serve experience your food is as much a matter of what is going on in their minds as it is a matter of what they put in their mouths.

However, it is not just the colour and other visual properties of food that set our expectations.

What's in a name?

Imagine yourself in a fancy restaurant, scanning the menu for something to eat. You already know that you want fish, but which one? Now, let's suppose you came across Patagonian toothfish. Would you order it? No, I didn't think so. Nor, for that matter, did anyone else. Sales of this veritable 'monster of the deep' had been disappointing for years. No matter how chefs prepared it, diners just turned their noses up and chose something else instead. Their eyes would continue scanning down the menu, looking for something that sounded, how shall I put it, a little more enticing.

Would the response be different, do you think, if they were to come across Chilean sea bass on the menu? It certainly sounds a lot more appealing, doesn't it? The thing is, though, that these two names refer to one and the same fish! Sales of this currently sustainable fish have increased by well over 1,000% – yes, that's three zeros – in a number of markets around the world (including North America, the UK and Australia). The trick was simple: just change the name. This is one of the most impressive examples of 'nudging by naming', as the behavioural economists like to call it. In fact, in no time at all, this fish started appearing on the menus of all the best restaurants, a trend that, even today, shows no signs of letting up. Once again, it is what is in the diner's mind, and the associations they make with different labels or descriptions that are crucial here.

The frozen crab bisque/smoked salmon ice cream and Patagonian toothfish cases are exceptional: they have, in fact, been chosen to make a particular point – about the importance of naming to our experience of food. Nevertheless, look around and one finds many everyday examples demonstrating much the same point. Have you ever wondered, for instance, why golden rainbow trout is so much more popular than regular brown trout? The traditionally trained chef's mind may immediately start to ponder differences in taste or texture, or perhaps to consider how the fish was dispatched. But why stop there? When was the last time you ate an *ugli* fruit (the result of the hybridization of a grapefruit – or pomelo – an orange and a tangerine)? Exactly. You have to wonder how much more popular this member of the citrus family could be had it been given another name. The decline in popularity of everything from faggots to pollack and Spotted Dick in recent years can, at least in part, be put down to their unfortunate names.

Great expectations

Some of you may already be wondering whether you could use the same naming 'tricks' to enhance people's perception of whatever food or beverage product you happen to be serving. Unfortunately, though, I very much doubt that you will be able to increase sales of most everyday foods by anything like as much as the Patagonian toothfish – sorry, Chilean seabass – example might lead one to believe. Nor, unless you have had your head very firmly stuck in the modernist cookbooks, will the colours of the dishes you prepare at home give as misleading an impression of the actual taste or flavour to come as the pinkish-red hue of Heston's frozen treat undoubtedly did. No one will, I presume, get the wrong end of the stick on seeing whatever you might be thinking of serving at your next dinner party. The colours of the foods we prepare normally give a pretty reliable indicator of the probable tasting experience. It is mostly in the modernist restaurant or when in parts foreign that things start to go awry. So relax!

Getting the name and/or description of a dish right is definitely worth investing some time in, even for those of you cooking at home. Take the following examples: simply referring to a pasta salad as a salad with pasta (i.e., just reversing the order in which the same words appear) makes people think of the dish as being that little bit healthier. And adding more descriptive elements, as when a restaurateur describes a dish as 'Neapolitan pasta with crispy fresh organic garden salad', is likely to lead to the number of positive comments that a dish garners increasing.

The topic of expectation management is just as important in the setting of the supermarket. Why else, after all, have the supermarkets started to create *phoney* farms to use in the labelling of their food packaging? Here I am thinking of farms like Rosedene and Nightingale; these names may well conjure up images of some rural idyll, but they do not actually exist. So why are the supermarkets doing it? Well, it turns out that we will pay more for exactly the same food, let's say a ploughman's sandwich, if we are told that the cheese inside was produced by farmer John Biggs, from Duxfield Farms in Cumbria. Obviously neither you nor I have any idea what this particular farmer's cheese tastes like, because I just made him up. And yet this kind of description adds value to the food offering or, in marketing-speak, it increases the consumer's willingness to pay. It may even make your sandwich taste different, perhaps better, as a result. These, then, are precisely the sorts of experiments that the gastrophysicist is interested in conducting, and the kind of results they want to share.

Others, though, have used the naming of the dish as an opportunity to capture people's attention. Heston Blumenthal received a phenomenal amount of press when he decided to call one of his new dishes 'Snail Porridge'; Had he given this dish a French name ('*Escargots à la Something*'), no one would have batted an eyelid; the dish would probably have tasted much more authentically French too. Over in Bror in Denmark, two ex-Noma chefs have decided to call one of their dishes simply 'Balls'. They come to the table breadcrumbed, fried to a reddish brown and dusted with sea salt. Delicious, apparently.

Paul Pairet, the chef at Ultraviolet, a multisensory experiential restaurant in Shanghai, has this to say on his restaurant's website: 'What is the "psycho taste"? The psycho taste is everything about the taste but the taste. It is the expectation and the memory, the before and the after, the mind over the palate. It is all the factors that influence our perception of taste.' So here's another of the world's top chefs explicitly recognizing the importance of 'the everything else' to the mind-blowing dining experiences that he provides.

Of course, we do not just have expectations about the taste and flavour of food and drink, and how much we will like it. We also have expectations about the kinds of food served by specific chefs or in specific venues; the same dish will taste very different to us as a function of whether it is served in a modernist restaurant, at your friend's house or up in an aeroplane. And then there is the anticipation, the booking of your meal. This is undoubtedly all part of pleasure too. You know, finding a great restaurant, even getting there, in some cases. Believe it or not, some chefs, with their minds squarely on the design of the experience, even consider how the diners will arrive at their restaurant. Just take Mugaritz, in Spain. As chef Andoni says: 'Mugaritz is not only the restaurant but also the road leading up to it, the countryside that you can see from the car and that, bend after bend, stokes the anticipation of everyone who visits us. Mugaritz is also its setting.'[2]

Or take Fäviken, the restaurant set in the wilds of the Swedish countryside. No one will doubt your credentials as a proper gastro-tourist if you manage to make your way to this remote location! The approach to El Celler de Can Roca, consistently rated first or second in lists of the world's top restaurants, is also chosen to discombobulate diners, situated as it is at the far side of an industrial park in Girona. So, should you be inviting any friends from afar to visit for dinner, be sure to recommend the scenic route.

'Tell me what you eat, and I will tell you what you are.' So said Jean Anthelme Brillat-Savarin in his much-quoted classic text *The Physiology of Taste*, first published back in the 1820s. Perhaps, but I would be tempted to put it rather differently: 'Tell me what a person expects to eat, and I'll tell you what they taste. I'll also estimate how much they'll enjoy the experience.' Expectations are key. Rare, after all, is the occasion on which we put something into our mouth without having first being informed, or having at least made a prediction, about what it is and whether or not we are going to like it. Our response to food – both the decision about what we choose to buy, order or eat, and what we think about it once we do – is nearly always affected by our beliefs (our expectations, in other words). It is the latter that subsequently anchor, and hence disproportionately influence, our tasting experience.

Do pricing, branding, naming and labelling influence taste?

Typically, we are aware of the brand and/or price of whatever it is that we eat and drink. On many occasions, the food will also be accompanied by some form of label or description. Such product-extrinsic cues, as they are known, all exert a profound influence over what people say about the taste, flavour and/or aroma of a food, not to mention how much they enjoy it. While we have known for years that pricing, branding and other kinds of product description can influence what people say about food and drink, until recently we had no real idea whether and how such factors affected the way in which the brain processes taste.

However, the latest neurogastronomy research demonstrates that the changes in brain activity resulting from the provision of such information can be dramatic. Differences are seen both in terms of the network of brain areas that are activated and the amount of activation that is seen there. What is more, these effects have, on occasion, been shown to affect the neural activity at some of the earliest (i.e., primary) sensory areas in the human brain. For example, in what has become one of the classic studies of branding, people had their brains scanned while one of two famous colas was periodically squirted into their mouth. Different patterns of brain activation were seen depending on which brand the participants thought that they were tasting.[3] The fact that branding has such a marked effect on flavour perception presumably helps to explain why blind taste tests are such a common feature of commercial product testing. There is, though, a question about what such tests actually tell us. Think about it for a moment. How often do you put something in your mouth without knowing what it is that you are tasting? While it may be a worthwhile exercise when it comes to the detection of flaws in food and beverage products, I suspect we should be doing more of our testing in the presence of all the other cues that normally go along with consumption. In that way, we will stand a much better chance of re-creating the more naturalistic conditions of everyday life.

Does food and drink taste better if you pay more for it? Not always, for sure, but more often than not. In support of such an intuition, neuroscientists in California investigated what happened in the brains of social wine drinkers (aka students) when given different, and sometimes misleading, information about the price of a red wine. A $5 bottle of wine was either correctly described or else mislabelled as a $45 bottle; meanwhile, a $90 bottle of wine was presented as costing either $10 or $90, and a third was correctly labelled as costing $35 a bottle. The price was displayed on a monitor whenever a small amount of the wine was squirted into the participant's mouth. In some trials, participants had to rate the intensity of the wine's taste, whilst in other trials they judged its pleasantness.

Everyone reported liking the expensive wine more than the cheap wine. Crucially, analysis of the brain scans revealed increases in blood flow in the reward centre of the brain associated with the price cue (see Figure 1.1). Telling people that the wine was more expensive (regardless of what wine they were actually tasting) led to an increase

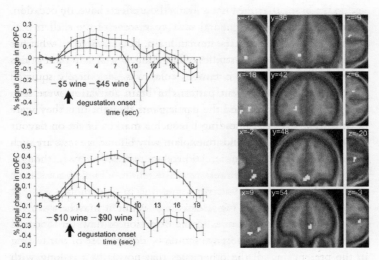

Figure 1.1. Images showing the percentage signal change in brain activation in the medial orbitofrontal cortex (mOFC; the brain's reward centre) over time (seconds are indicated on the *x* axis) as a function of the price associated with a wine that the participants in the scanner are tasting.

in activation in the medial orbitofrontal cortex (mOFC), a small part of the brain located just behind the eyes. By contrast, no change in blood flow was observed in the primary taste cortex, the part of the brain that processes the sensory-discriminative attributes of taste (e.g., when judging how sweet, sour, etc. something is). Intriguingly, though, when the same wines were presented eight weeks later, now without any indication as to their price (and away from the confines of the brain scanner), no significant differences in pleasantness were reported. And the latest evidence suggests that the effects of misleading pricing may work better in the mid-price range. So I am afraid that no matter what you say, you have little chance of convincing people that the Two-buck Chuck (i.e., cheap plonk) you are serving them is premier cru.

Imagine being given a clear solution to drink. You are told either that you are about to taste something very bitter or else something that is much less bitter. Should you happen to be lying in a brain scanner at the time, changes are likely to be seen in some of the earliest sites in the brain after the taste and smell signals are initially coded by the sensory receptors. In particular, researchers have shown that activity in the middle and posterior insula, an area deep in the cerebral cortex, can be modulated by the verbal description people had been given regarding the intensity of the to-be-delivered taste. The response in the brain's reward centre, the OFC, also varies systematically as a function of people's expectations concerning the drink's bitterness. Elsewhere, researchers have tried verbally describing an odour as 'smelly cheese' and found that people rated it as more pleasant than when exactly the same odour was labelled as 'sweaty socks' instead. Once again, the brain's response was modified as a result of the provision of a product-extrinsic cue.

While such neuroimaging results are undoubtedly fascinating, it is perhaps worth bearing in mind here just how unnatural the situation in which the participants find themselves really is. How often, after all, do you go out on a Friday night and find yourself lying flat on your back, inserted several feet into a narrow tube, with your head clamped still. The latter precaution is needed in the brain scanner in order to minimize any head movements that can make it difficult to analyse the neuroimaging data. And that is not all. You will have a tube held between your teeth as a few millilitres of wine is periodically squirted into your mouth. You are told to evaluate its taste, without swallowing. Eventually you are allowed to swallow, then your mouth is washed out with artificial saliva (yes, really). Then the whole process starts again.

People's beliefs about the origins of their food also impact on how they perceive taste. For instance, in a recent study illustrating this point, students from the US were given identical samples of meat (e.g., beef jerky or ham) and told either that it was factory farmed or that it was free range. Those who were told that the meat was factory farmed rated it as tasting less pleasant, saltier and greasier. What is more, the students ate less of it too, and said that they would have been willing to pay less for the meat. Crucially, the same pattern of results was obtained across three separate studies. One finds that describing a food as organic or free range has much the same effect – despite the fact that in blind taste tests consumers mostly cannot tell the difference. So what this means in practice is that if you shell out for some organic, free-range, hand-fed food, you should be sure to let your guests know its provenance if you want them to be able to taste the difference.

One of the many challenges facing food and drink companies in this area is that even though they may be making real and sustained progress in terms of reducing the less healthy ingredients in their branded products, they are often best advised not to state 'low fat' or 'reduced sugar' on their labels, because doing so is likely to cause the consumer to say that it tastes different. Keep quiet about it, and they may not detect that anything has changed. Health by stealth, that's the way to do it! It is worth stressing here that the interests of the food and beverage industry tend to be quite different from those of the modernist chef. The latter is trying to create unusual, surprising and sometimes spectacular results. The majority of the diners at the top restaurants tend not to care too much about the health/nutritional content of the meals they are served (given that it is likely to be a one-off occasion). Rather, they want surprise and novelty. The former, by contrast, are typically more interested in trying to keep their successful branded products tasting the same to consumers as they always did, whilst gradually making their products less unhealthy.

Once you understand just how important naming, labelling, branding and pricing can be, you might start to wonder how much, if anything, is actually happening at the level of the taste buds. Ultimately it is the interaction between what is in the mouth and what is in the mind that determines what the final tasting experience is like, and how much we enjoy it. Master both the food and the gastrophysics and you'll be in a good place to impress, whoever you are and whomever you are cooking for.

Taste worlds

Tell me, what does coriander (or cilantro) taste like to you? Do you love it or loathe it? The majority of people, it has to be said, like its fresh, fragrant or citrusy characteristics. Others, by contrast, are convinced that it tastes soapy (some even describe spinach as soapy too). It reminds them of dirt, bugs or mould, they say. Those in the latter camp will typically avoid *any* food containing what John Gerard, writing back in 1597, called a 'very stinking herbe' with leaves of 'venemous quality'.[4] So who is correct? What does coriander *really* taste of?

Both sides are right, though more of the population fall into the former category. Most of us – 80% or more – are likers, the exact figure depending on the ethno-cultural group tested. Are those on the soapy side of the spectrum simply unable to detect one of the many compounds that make up the distinctive flavour of coriander? Or perhaps those on the citrusy side are anosmic to something ('anosmia' being the technical name for being unable to smell some volatile chemical or other). No one knows for sure! What is more, there is even uncertainty about whether that soapy sensation should itself be characterized as a taste, an aroma or something else entirely. Whatever it is, it doesn't seem to fit any of the commonly recognized basic tastes.

Though this may be more of a topic for the next chapter, it is worth noting here that something like one in every two people can't smell androstenone, an odorous steroid derived from testosterone. They are anosmic to this particular volatile organic molecule. Meanwhile 35% of the population find that it has a very powerful – and deeply unpleasant – stale, sweaty, urine smell. (This is the reason why male pigs are castrated, i.e., to minimize the unpleasant aroma known as 'boar taint'.) Worse still, the individuals in this group tend to be exquisitely sensitive to this compound; some can detect it at concentrations of less than 200 parts per trillion. The remaining 15% or so of the population, well, they say that it smells sweetly floral, musky and/or woody. Some people (like me) experience the smell simply as chemical-like. Same molecule: completely different experience!

The prevalence of these genetic differences in the worlds of taste/flavour perception varies by region and culture. So, if you had to guess, in which part of the world do you think the likelihood of people perceiving the urinous note in their uncastrated pork meat would be highest? I have heard that it's the Middle East – i.e., exactly the place where religion bans pork as a legitimate source of food. Just mere coincidence, you think? Seems unlikely, doesn't it?

Coriander and androstenone are just the tip of the iceberg as far as genetically determined differences are concerned. That is to say, every one of us is anosmic to some number of compounds, many of which are associated with food. So, for instance, our sensitivity to isovaleric acid (a distinctive sweaty note in cheese), ß-ionone (a pleasant floral note added to many food and drink products; think of the fragrance of violets), isobutyraldehyde (which smells of malt) and cis-3-hexen-1-ol (which gives food and drink a grassy note) all show a significant degree of genetic variation, and roughly 1% of the population are unable to smell vanilla. What this means, in practice, is that there are some pretty profound individual differences in people's ability to perceive these compounds.

Who knows, then, how many of the disputes between wine experts can be put down to such genetic variability? Just take the famous disagreement between Robert M. Parker Jnr, the influential American wine critic, and the British Master of Wine Jancis Robinson regarding the 2003 Château Pavie. The former absolutely loved this wine, whereas the latter slammed it, giving the *en primeur* wine a score of 12/20. Robinson had the following to say: 'Completely unappetising overripe aromas. Why? Porty sweet. Port is best from the Douro not St Emilion. Ridiculous wine more reminiscent of a late-harvest Zinfandel than a red Bordeaux with its unappetising green notes.' Parker responded by saying that the Pavie 'does not taste at all (for my palate) as described by Jancis'. So were these two international experts tasting the same wine differently? Did they perceive the same attributes, which one writer appreciated and the other disliked? Or did the wine really taste different to the two star wine writers?[5]

I myself am totally anosmic to tri-chloro-anisol (TCA for short), the chemical that gives rise to cork taint in wine. This form of 'smell blindness' is something that my wine colleagues find most amusing, as I am sure you can imagine. When a corked bottle comes to the table, they will order a replacement and put two glasses of the same wine down in front of me, one from either bottle. I will normally find them to be identical, whereas my friends will not be able to drink from one of the two glasses. Once again, TCA is one of those chemicals to which people show huge differences in terms of sensitivity. I normally have the last laugh, though, since when the uncorked wine runs out, there is still plenty more wine left that only I enjoy!

What I'm hinting at is that we all live in very different taste worlds (see Figure 1.2). Some people are able to detect bitterness in food and drink where others taste nothing (the former group are commonly referred to as supertasters). Supertasters may have as many as sixteen times more papillae on the front of their tongue as others (known as non-tasters). Not only do people vary in terms of their sensitivity to bitterness but also – to a less pronounced degree – in terms of their perception of saltiness, sweetness, sourness and oral texture. Taster status, like odour sensitivity, is largely heritable (i.e., genetically determined). In fact, back in the 1930s, scientists were thinking of using this taste test as a paternity test. And beyond these individual differences in sensitivity to the basic tastes, we all vary quite markedly in terms of our

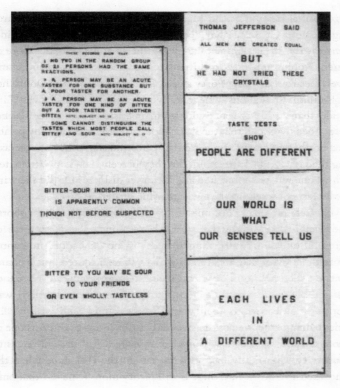

Figure 1.2. One of the original posters from a public demonstration of the different taste worlds in which we live – from the 1931 New Orleans American Association for the Advancement of Science Meeting.

hedonic responses too. So, for example, there are those who are sweet likers, whereas others (including myself) are best classified as being more ambivalent about sweetness.

But why should bitterness be the taste for which individual differences are most pronounced? Why are the individual differences not so apparent for the salt, sweet or sour tastes? It is likely that individual differences in sensitivity to bitterness may have been especially important for our ancestors. In times of plenty, the supertasters would have had a competitive advantage, since they would be unlikely to ingest something bitter and hence potentially poisonous. By contrast, in lean times, it would have been the non-tasters who had a slight competitive advantage since they would have been more likely to ingest those bitter foodstuffs that happened not to be poisonous and hence less likely to starve to death. It is a little harder to make such an argument for the other tastes.

However, a liking for bitter-tasting foods (associated with super-taster status) also correlates with psychopathic tendencies! Or, as the authors of one recent study put it: 'General bitter taste preferences emerged as a robust predictor for Machiavellianism, psychopathy, narcissism and everyday sadism.'[6] Though, of course, it is important to note that correlation is not causation – you are not necessarily a psychopath should you be one of those who likes bitter-tasting food and drink. Intriguingly, the latest research shows that tasting something bitter can give rise to increased hostility. By contrast, tasting something sweet can apparently make you feel more romantic and increase the likelihood of you agreeing to go on a date. Even more remarkably, those who are thinking about love will rate water as tasting sweet. Meanwhile, men whose hockey team had won rated a lemon-lime sorbet as tasting sweeter than did those supporting the losing side. And going a step further, marketing professor Baba Shiv and his colleagues in California have reported that handling large amounts of money can change people's taste thresholds. Once again, the sense of taste turns out to be so much more than merely a matter of taste.

Some global food companies have already taken advantage of this distinction by launching two versions of a particular product into the marketplace, one targeted at supertasters, another for all the non-tasters out there. Not that it will say this on the label; the company will just let the market segment itself. Remember: taster status runs in families. As it happens, my mother, brother, sister and nieces are all supertasters, whereas my father can taste none of the bitterness in broccoli that the rest of the family can. This, I believe, helps to explain why my father would always make us kids eat all of our vegetables before leaving the dinner table. He never understood, or so we like to think, how terrible these green vegetables tasted to the rest of us. If only we had known then that people live in different taste worlds.

'There is more to taste'*

Taste is crucial to our survival. In a way, one might think of it as the most important of our senses – helping us to distinguish between that which is nutritious and that which may be poisonous. And yet, on closer inspection, it turns out not to be so important, at least not in terms of perception. You get a sense of this from looking at how much of the cerebral cortex is given over to each of the senses. While more than half of the brain is involved in processing what we see, only something like 1% of the cortex is directly involved in taste perception. The reason for this is that our brains pick up on the statistical regularities of the environment, and so we learn to predict the likely taste and nutritional properties of potential foodstuffs on the basis of other sensory cues, such as colour and smell. We may learn, for example, to expect pinkish-red foods to be sweet. This allows us to assess the likely consequences of ingesting a whole host of different foods without necessarily having to stick them into our mouths first in order to determine what they taste like.

Ultimately, if you know about the expectations set up by the other senses, then you are in a much better position to modify people's perception of taste. It may even help all those exasperated parents out there wondering how to get their offspring to eat more vegetables. So, no matter how you define (or think about) taste, it is clear that the other senses play a far bigger role in determining what we think we are tasting and how much we enjoy the experience than we generally realize. So, in closing, let me introduce you to Eleanor Freeman, senior snack inventor at online health food company Graze. Her taste buds have been insured for £3 million. Gennaro Pelliccia, an Italian coffee master for the Costa coffee chain in the UK, has his

---

* This is the strapline from a recent advert for Lavazza coffee.

insured for £10 million, while the taste buds of Hayleigh Curtis, who works for Cadbury chocolate, are insured for a measly £1 million. To me, though, this just sounds like a headline-grabbing stunt, for as we will see in the next chapter, what a top taster really needs to worry about is their nose.

2.   Smell

Think about the last time you had a head cold and your nose was blocked. Food and drink didn't have much of a taste, did they? Ever wonder why that is? What is missing in such circumstances is not taste – trust me, your taste buds are working just fine – but rather aroma. Assuming that you don't have a cold at the moment, try holding your nose pinched tightly closed, and get a friend to feed you something without letting you know what it is. Unless they pick something really pungent (and if they do that, maybe they aren't such a good friend after all), you will most likely have very little idea of what you are tasting – onion or apple, red wine or cold coffee. These pairings are surprisingly hard to tell apart without a functioning sense of smell.★

It is important to distinguish here between the two different ways in which we smell. There is the 'orthonasal' route: when we sniff external aromas from the environment. And there is 'retronasal' smell: when volatile aromatic odour molecules are pulsed out of the back of the mouth into the back of the nose whenever we swallow while eating and drinking. The orthonasal sniffing of food aromas is especially important because it allows us or, rather, our brains, to form the rich flavour expectations concerning both what the experience of tasting will be like and how much we expect to enjoy it. But it is the retronasal perception of aroma, on swallowing, that really provides our tasting experiences with their rich variety and interest. Most of the time, though, we remain acutely unaware of how much of the information that we think we are tasting via the tongue

---

★ Or, for a mischievous version (borrowed from Sam Bompas, one colourful half of the jelly-mongers extraordinaire Bompas & Parr), why not chop up some cabbage, boil it in water and pour the liquid into a teapot, then surprise your friends when they let go of their noses with something that looks like tea but has a most unsavoury aroma!

comes in via the retronasal olfactory route. This is, in large part, because food aromas are experienced as if coming from the mouth – as if being sensed by the tongue itself. This strange phenomenon goes by the name of 'oral referral'.

To illustrate the point, try eating a jelly bean with your nose closed between thumb and forefinger. What can you taste? You will most likely experience sweetness, maybe some sourness and, who knows, perhaps a hint of spiciness too (at least if you get a cinnamon one). Then, after a few bites, let go of your nostrils. You will suddenly get the fruity flavour, orange or cherry, etc. But you will experience that flavour as coming from your mouth, not from your nostrils. That is oral referral in action, the mislocalization of aroma to the mouth!

Does vanilla smell sweet to you?

The answer for most people is a definite 'yes'. People give exactly the same answer when it comes to the aroma of caramel and strawberries too. Now, this is confusing, right? After all, didn't I just say in the last chapter that 'sweet' was a taste descriptor? So how can an aroma be said to *smell* sweet? Some have argued, I think wrongly, that this is a kind of synaesthesia. Interestingly, food companies add vanilla flavouring to ice cream to bring out the sweetness. They do this because at very cold temperatures your taste buds don't work so well, and hence you can no longer taste sweetness – but you can still smell it. Surely you have had the experience of drinking a warm glass of cola by mistake? Doesn't it suddenly taste sickly sweet? The composition of the drink itself hasn't changed, but the signals that your taste buds send through to your brain have changed as a function of the drink's temperature. Since the drink is usually served cold, the manufacturer has added some sweetness through the nose. Confused? You should be.

In the other direction, i.e., when it comes to taste's influence on aroma and flavour perception, things are very different. One of the classic studies in this area involved people tasting a solution whose sweetness had been carefully calibrated to be just below the level of awareness (in other words, it tasted like plain water). Nevertheless, when people held a small amount of that subjectively tasteless liquid in their mouth, their ability to detect the cherry-almond aroma in another drink that they were smelling suddenly increased dramatically. Importantly, however, further research showed that the taste had to be *congruent* with the smell in order for this effect to occur. Adding a sub-threshold dose of monosodium glutamate into the mouths of Western participants didn't have the same effect. However, the response may well be different in Japanese consumers. In other words, the research suggests that while everyone's brain uses the same rules to combine the senses, which particular combination of tastes and smells gives rise to the enhancement or suppression of flavour depends on the food culture that person has grown up in.

The amazing thing is how quickly this kind of learning happens, and what's more it continues to occur throughout our lifetimes. Take a novel odorant, the smell of water chestnut (as in one study conducted on Australian adults a few years ago). Then, simply pair it with either a sweet or bitter tastant in the mouth. Believe it or not, after no more than three co-exposures, the smell starts to take on the appropriate taste qualities. More remarkably still, this can occur even when the tastant is presented at a level below perceptual awareness.

Have you ever noticed how freshly ground coffee often smells wonderful, yet when you come to taste it the flavour can be a little disappointing? The same thing in reverse occurs if you take a ripe French cheese. It may well smell like the inside of a jock's training shoe (please excuse the metaphor), yet if you can manage to put some in your mouth, the pleasurable experience that follows is often sublime. What is happening? These changes in our hedonic ratings – basically, how much we like something – illustrate the distinction between our two ways of smelling: orthonasal (when breathing in) and retronasal (when breathing out through the back of the nose). Normally, we are remarkably good at predicting the likely retronasal flavour of a food based on nothing more than an orthonasal sniff. So good, in fact, that we simply don't know we're doing it.

Scenting the scene

Look around the world of high-end modernist cuisine and molecular mixology and one sees an increasing use of scene-setting scents and mood-inducing aromas. These are being added to dishes, to drinks, to the dining table and even, on occasion, to the entire dining room (especially in those situations where the chef has the luxury of serving a single sitting, with everyone eating the same course at the same time). The aim in many cases is to create a particular atmosphere or mood, or else to trigger a specific memory in the mind of the guest, no matter what they happen to be consuming. For example, Heston Blumenthal serves a moss-scented 'Jelly of Quail with Langoustine Cream and Oak Moss' dish at his flagship restaurant, The Fat Duck. Steaming scented vapour pours out from the moss-scented carpet placed in the centre of the table (see Figure 2.1). No one, I presume, starts salivating at the thought of a mouthful of the green stuff and yet the theatrical use of scent definitely helps transport the diner to

Figure 2.1. One of the fragrant dishes served at The Fat Duck. The scent of moss covers the table and fills the diners' nostrils.

another place and in so doing enhances their experience of the dish. At Alinea in Chicago, hot water is poured over a bowl of flowers when the 'Wild Turbot, Shellfish, Water Chestnuts, and Hyacinth Vapor' dish is served. The chef, Grant Achatz, is also famous for his pheasant, served with shallot, and cider gel, which is accompanied by burning oak leaves. The idea here is to use scent to trigger happy memories of an autumnal day from childhood.

Of course, one has to be careful not to overuse scene-setting scents. One diner, writing on TripAdvisor about their experience at The Fat Duck, stated: 'The final dish was "Going to Bed" ["Counting Sheep"] and I think was meant to be reminiscent of being a baby, but the smell of baby talc was overwhelming and that's not an aroma one necessarily wants while eating.' While this was certainly not my recollection of the dish, the quote does highlight the potential danger for anyone trying to use a background scent that may compete with the foreground aromas of the food itself. The challenge is, in part, made worse by the fact that even ambient smells can be mislocalized to the oral cavity and experienced as tastes or flavours in the mouth if one is not careful. (Yes, oral referral strikes again!)

Fortunately, the gastrophysicist has a few tips up his sleeve to show the modernist chef/molecular mixologist how best to convince their customer's brain to segregate the background environmental smell from the foreground aroma of the food and drink (assuming, of course, that that is what the chef is trying to do). Ensuring that the various smells are first encountered at different points in time will help here, as this will make it easier for the customer's brain to localize the background aroma in something other than the food or drink. And presumably Achatz keeps his oak leaves and hyacinths very visible in order to do the same. Crucially, by using such an approach, the perceived source of the scent is likely going to be correctly localized in something other than the food.

I would like you to imagine that a sugar cube is soaked in a few drops of rose oil and placed into a glass of champagne. Imagine the drink sitting in front of you, effervescing gently. The fragrant smell of an English rose garden emanating from the glass surrounds you. Before you know it, you find yourself being transported to a pleasant scent-infused summer afternoon somewhere in your memory. This is exactly what top mixologist Tony Conigliaro, of 69 Colbrooke Row, wants you to experience.

Conigliaro is using scent to prime positive memories and associations. One of the specific advantages of this approach is that smell has a much closer, more direct connection with the emotional and memory circuits in our brain than any of our other senses. It turns out that the olfactory receptors in our nose are actually an extension of our brain. In fact, it is only a couple of synapses from the cells in the olfactory epithelium lining the inside of the nose through to the limbic system, the part of the brain that controls our emotions. By contrast, information from the other senses has a much longer path to travel through the brain before it hits the emotion centres, and hence it can be more easily filtered out. The challenge, though, for a multi-course, smell-enhanced tasting menu is how to clear one fragrance out before the next course arrives, one of the key practical problems that eventually scuppered early attempts at scented cinema – remember Smell-O-Vision, anyone?

If smell is indeed such an important part of what we taste, and if it is such an effective means of triggering our moods, emotions and memories, then any one of the innovative approaches mentioned thus far in this chapter really makes sense from the gastrophysics perspective. However, those of you who are not lucky enough to eat or drink at one of these gastronomic hotspots might be thinking: how exactly am *I* supposed to use this knowledge? In the next section, I want to share some of the intriguing ways in which we will all be exposed to this new world of augmented flavours in the coming years. For where the modernist chefs, molecular mixologists and culinary designers lead, you can be sure that food and beverage manufacturers are never far behind.

Making sense of smell

Given how important smell is to our enjoyment of food and drink, it is surprising to realize quite how many of our everyday food – and especially beverage – experiences are not optimized to deliver the best orthonasal aroma hit. Perhaps the best (or should that be worst?) example of poor olfactory design are those plastic lids that are routinely placed over millions of paper cups of hot coffee. While these lids undoubtedly allow you to drink without having to worry about spillage, what they singularly fail to do is to allow the drinker to appreciate the orthonasal aroma of the cups' contents by sniffing them. This is especially unfortunate in the case of a freshly ground cup of coffee, given that it is one of the most universally liked smells. Much the same problem occurs whenever we drink directly from a bottle or can (uncouth as it may be!). Once again, it is the orthonasal olfactory hit that is mostly missing from the experience. We can either sniff the contents or we can drink them, but there is simply no way that we can do both at the same time, no matter how hard we try. And drinking through a straw – well, that is even worse!

So, having identified the problem, what can be done about it? In terms of design, there are a number of simple solutions out there. They include the reshaping of the lid, or adding a second opening to allow the coffee (or tea) lover to sniff the aroma of their favourite beverage while sipping. This is part of the innovative solution incorporated into the ergonomic lid introduced by Viora Ltd. Their novel design allows the consumer to smell their coffee without having to take the lid off. Common sense, really. But if that's the case, you have to ask yourself why it took so long for someone to come up with this solution. My suspicion is that it is all down to oral referral again. It's not obvious that retronasal smell is involved in tasting, and hence no one bothers to factor it into their designs.

Another intriguing solution comes from Crown Packaging. The company designed a can (see Figure 2.2) with a top that lifts off completely, allowing the thirsty consumer to see and, perhaps more importantly, sniff the contents more easily than when drinking from a traditional can (i.e., one with a small tear-shaped opening).

At the opposite extreme to the traditional lid, bottle or can that resolutely prevents the consumer from orthonasally enjoying the aroma of whatever it is that they are drinking, let's take the humble pint. Back in the day, when all lager beers used to taste the same, the lack of any protected headspace over the drink in the glass probably

Figure 2.2. Two examples of enhanced olfactory design: the Viora lid (*left*) and Crown's 360End™ can (*right*).

didn't matter much. However, given the revolution in craft beer in recent decades, there are now a host of drinks out there that many of us are willing to pay a hefty premium for (because, for a change, they really do taste of something). The problem with the traditional pint glass filled (as it always is) to the rim is the lack of any protected headspace over the beer in the glass. This, in turn, means that there is no way of concentrating the drink's aromas. So, assuming we take it as read that a more intense smell is a good thing, perhaps we should be thinking again about the design of the glass in which so many pints of beer are served every day as well. So what, one might ask, would the gastrophysicist recommend here?

You might start by considering what happens in the world of wine. After all, there is always around ten times more research into wine than into anything else that we might be tempted to drink (presumably because researchers like drinking wine). Firstly, notice how wine glasses are never filled to the rim. Somebody out there certainly believes, rightly or wrongly, that the empty headspace over the drink is important. It is meant to help preserve the aroma and bouquet of the contents of the glass, so as to delight the taster's nostrils. In fact, the better the wine, the larger the proportion of the glass that is left empty, or at least so it would seem.

One might, of course, argue that having a full pint glass doesn't really matter; after all, as soon as the thirsty drinker has taken a few draughts from their pint, won't they have created an aromatic head-space over the drink anyway, so why worry? Bear in mind, though, that more often than not the initial sniff will set up expectations about what's coming next. It is these expectations that end up anchoring, and hence disproportionately influencing, the tasting experience that follows. Isn't that first mouthful so much more important (not to mention more enjoyable) than any of the swigs you take in the middle of your pint? And both have got to be better than the last mouthful, when all you are left with are the warm flat dregs in the bottom of the glass. So, if we value the flavour and aroma – as we should – then

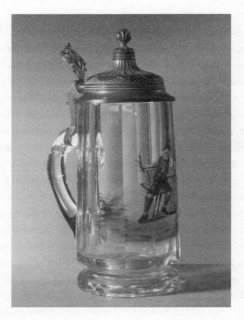

Figure 2.3. Lidded stein glass: an early example of intelligent olfactory design?

maybe we should all make sure a little room is left in the glass when the beer is first served.

Of course, the danger is that the average beer drinker has become so accustomed to seeing their glass filled to the brim that they might feel short-changed were they to be given anything less. An alternative solution to this olfactory problem might well simply be to dust off those old stein glasses that once came with their very own lid attached (see Figure 2.3). The purpose, at least according to one early commentator, writing back in 1886, was to protect the gases released from the beer's surface.[1] I like to think of this as clever olfactory design from 130 years ago!

How can the delivery of flavour be enhanced?

Have you ever noticed how little aroma there is in the public areas in airports? Walk into a railway station or book store and your nostrils will almost certainly be assaulted by the smell of coffee. Airports, by contrast, appear to be olfactorily neutral spaces. That said, next time you pass through London Heathrow Airport's Terminal 2, why not stop for a bite to eat at The Perfectionist's Café. If you do get the chance, then I'd recommend the fish and chips. What may surprise you about the dish is the use of an atomizer to dispense the aroma of vinegar. This, as we will see below, is just one of the ways in which creative individuals are starting to change how the more aromatic elements in a dish are delivered to the table.

Over the last couple of years, London-based chef Jozef Youssef has been experimenting with the atomized delivery of aroma in a number of the dishes he serves. During his 'Elements' dinners, for instance, the mossy-earthy smell of geosmin was sprayed over each and every diner's bowl of leek consommé, leek ash and goat's cheese cream. Meanwhile, in the chef's sell-out 'Synaesthesia' dining events, it was the scent of saffron (saffranel) that was sprayed over a butter-poached lobster with white miso velouté instead. Sounds simple, right? So, why don't you try this for yourself (i.e., 'atomizing' your guests) the next time you invite them over for dinner? You never know, do it right and they may just thank you for the way their senses are awakened and for how much better their food tastes as a result. All you need is a small, clean spray bottle in which to place your 'food perfume'.

It is F. T. Marinetti and the Italian Futurists, though, who really deserve the credit for first bringing atomizers to the dining table, back in the early decades of the twentieth century. Though they were more likely to spray perfume (smelling of carnations) into their diners' faces should any of them be foolish enough to look up from their plates while eating. Quite what effect this had on the multisensory tasting experience has sadly not been recorded for posterity! But the Futurists were much more about provocation than about delivering the best multisensory dining experience possible.

Today, inventive chefs and mixologists are using smoking guns to deliver the requisite aromas to the dishes and drinks they serve (see Figure 2.4). The dry-ice-based cloud pourer allows the creative flavourist to add concentrated aromas to a dish or drink in the form of a misty vapour that can be poured tableside, or at the bar, right in front of the wide-eyed and open-mouthed customer. Come on now, don't be afraid . . .

Figure 2.4. A smoking gun – the chef's best friend?

You might also not have realized that aromatic packaging has been in the marketplace for years. Just take the humble chocolate ice cream bar. While everyone loves the smell of chocolate, the real stuff lacks its delicious aroma when frozen. One manufacturer even tried adding a little synthetic chocolate aroma to the glue seal of the packaging to make up for the lost smell, so that when the customer ripped open the packaging they caught a whiff of chocolate smell which they, naturally enough, assumed must have come from the chocolate covering the ice cream itself. Not everyone uses scent-enabled packaging solutions, that's for sure. Truth be told, it can be tricky to deliver an authentic-smelling encapsulated chocolate aroma.

Then, of course, there are the reports of coffee companies injecting various aromatic substances (some, so the probably apocryphal stories go, extracted from the rear end of the skunk) into the headspace of their packaged coffee. This presumably helps to explain why the experience so many of us have on first opening a packet of coffee can be so great. Your nostrils get a sharp and pungent hit of what you undoubtedly thought was great-smelling freshly ground coffee. But why, you should ask, is the experience nearly always disappointing when the container is opened again subsequently?

Another intriguing example of how design can be used to modify flavour came with the launch of The Right Cup in 2016. This glass drinking vessel includes a colourful sleeve that gives off a fruity aroma. The apple-flavoured cup is bright green, the lemon yellow, and the orange – well, what else but orange-coloured? The idea of the cup is that the consumer can drink water from it and have a tasting experience that approximates to what might be expected on actually drinking the appropriate fruit juice, or at least fruit-flavoured water. I bet that the colour cue provided by the sleeve will turn out to be pretty important to the tasting experience. Along somewhat similar lines, in 2013, PepsiCo applied for a patent for the use of encapsulated aroma in their drinks packaging.[2] These scented capsules would only be broken, and the aroma released, when the consumer unscrewed the lid. The thinking was that a better aroma experience could be delivered by scenting the packaging rather than the product itself.

Canadian company Molecule-R sells an Aromafork kit: for

around US$50, you get a set of four metal forks, with a bag of circular blotting papers to insert into the end of the fork, and twenty small phials of different aromas intended to augment the flavour of food with each and every mouthful. I have yet to try The Right Cup, but my experience with the Aromafork is that unless one is very careful the aroma can all too easily end up striking one as synthetic. This was certainly the response of the guests on the BBC Radio 4 show *The Kitchen Cabinet* when I tried the scent-enabled fork on them. This is not to say that we can always distinguish synthetic from natural aromas; more often than not, we can't. The problem here is just that many of the aromas that are currently sold with the kit smell cheap and artificial, and most of us certainly don't like our food to taste artificial.

I remain to be convinced by Molecule-R's suggestion that their innovative fork is ideal for the home chef who has forgotten to add a particular ingredient to a dish. Its most beneficial use, as far as I can see, would be to replace some particularly expensive ingredient. I can, for example, imagine dribbling a few drops of quality truffle oil, say, on to my fork, thus delivering far more culinary pleasure than if the same amount of oil were simply to be drizzled haphazardly over the food itself. The atomized scent of saffranel could be used in much the same way – saffron being, it is said, gram for gram, more expensive than gold. So are you game to try this at home? Simply put a few drops of something fragrant on to the middle of a wooden spoon or fork before using it to serve your guests. You'll be guaranteed to deliver a dining experience with a difference!

Slow food this most definitely is not. But should the consumer realize that they can get a better experience, or else the same flavour, at a fraction of the price (at least in the case of truffle, saffron and other similarly expensive ingredients), then who knows, the Aromafork – or some more aesthetically pleasing successor – might just revolutionize how we eat in years to come.*

---

* Assuming, that is, that our desire for those pricey ingredients is really about the taste, and isn't just functioning as a Veblen good – i.e., as something used to signal our wealth to others.

Ultimately, the chances of such augmented approaches succeeding in the long term are going to depend on the delivery of quality scents at a reasonable price. Remember the perceived synthetic (rather than natural) nature of the scents in the Aromafork. Once the sniffer realizes that what they are sniffing doesn't originate from their food or drink but instead comes from the cutlery, glassware or packaging, they might well start to believe that it smells artificial. Rightly or wrongly, there are frequent scares about our exposure to synthetic flavours and scents in everything from fragranced candles through to processed foods. It is that belief or concern that the smell is artificial or 'chemical' – though, of course, *all* aromas are chemical – as much as the evidence before our nostrils that will, I predict, reduce hedonic ratings when sampling these products.

Have you noticed how often the modernist chef stresses, either explicitly or otherwise, the *natural* origins of their off-the-plate aromas? Be it the scent that is released when hot water is poured over the hyacinths at Alinea, or the recently deceased Homaro Cantu's use of fresh sprigs of herbs in the curly handles of his cutlery at Moto, also in Chicago. We will just have to wait and see how the consumer of tomorrow responds to this all-new world of olfactorily enhanced food and beverage packaging, glassware and cutlery. No doubt the food and beverage companies, together with the flavour houses, will take a page out of the chefs' and mixologists' book, and figure out how best to emphasize the naturalness of their novel olfactorily enhanced design solutions.

The olfactory dinner party

So far, we have focused on the use of scent to deliver both food and non-food aromas in a more effective manner, so as to enhance the multisensory flavour experience, or else to provoke a certain mood, memory or emotion. But can flavourful smells also be used to promote healthier eating behaviours? Go back to the 1930s and one finds the Italian Futurists (yes, them again) suggesting that 'in the ideal Futuristic meal, served dishes will be passed beneath the nose of the diner in order to excite his curiosity or to provide a suitable contrast, and such supplementary courses will not be eaten at all'.[3] The same notion makes an appearance in Evelyn Waugh's 1930 novel *Vile Bodies*: 'He lay back for a little in his bed thinking about the smells of food, of the greasy horror of fried fish and the deeply moving smell that came from it; of the intoxicating breath of bakeries and the dullness of buns . . . He planned dinners, of enchanting aromatic foods that should be carried under the nose, snuffed and then thrown to the dogs . . . endless dinners, in which one could alternate flavour with flavour from sunset to dawn without satiety, while one breathed great draughts of the bouquet of old brandy . . .'[4]

Smell (olfaction), then, is more important to our tasting experiences than any of us realize. That being the case, you might be thinking, why not simply enjoy the aromas of great-tasting food without the calories associated with actual consumption? That, at least, is one of the ideas behind the olfactory dinner party. However, you don't need a gastrophysicist to tell you that your cravings are never really going to be satisfied by smell alone.

And yet there are a growing number of companies out there who are starting to deliver food aroma as an end in itself. Just take the coffee inhaler: wherever you might be, you can get a caffeine hit without having to find a coffee shop. You can inhale chocolate aroma too. Culinary artists Sam Bompas and Harry Parr fit right in here, with their 'Alcoholic Architecture' installations. These UK-based jelly-mongers have experimented with a series of 'cloud bar' installations, where punters could spend fifteen minutes or so in a space infused with a gin & tonic mist. Then there is the Vaportini, a bit of kit that gently heats your drink and concentrates the aromas, so that you can just inhale them and supposedly enhance your experience as a result.

This all sounds intriguing enough, but I really don't see the idea of the olfactory dinner party gaining popularity any time soon. For I doubt that our brains can be truly satisfied without our actually consuming something. As Lockhart Steele, owner/creator of *Eater.com* puts it: 'Novelty is everything in a certain corner of the dining world, no matter how fleeting . . . Dining in the dark, dining without talking – all that's left is eating without eating.'[5] And as we will see in the 'Touch' chapter, oral-somatosensory (i.e., mouth touch) and gustatory (or taste) stimulation appear key to driving our brains towards satiation (i.e., fullness).

That is not to say that enhancing the delivery of food aromas while *actually* eating and drinking (what I am calling 'augmented flavour') might not be a good idea. The female participants in one recent laboratory study became fuller sooner following the enhanced delivery of food aroma while they were eating a tomato soup. In this case, simply ramping up the olfactory component of the dish reduced people's consumption by almost 10%.[6] Thus, all of us might be able to reach satiety sooner if only we knew how to stimulate our senses more effectively. Such findings support the idea that augmenting the orthonasal aroma of food and drink could (say, through food and beverage packaging or via smelly cutlery) lead to greater enjoyment, not to mention possibly smaller waistlines.

Scents and sensibility

Have you been to a Hilton Doubletree hotel recently? If so, you will no doubt be familiar with the deliciously sweet and aromatic cookie scent that fills the lobby at check-in. Smile at the person behind the counter and they are likely to give you a freshly baked cookie. Once again, we see a desirable food aroma being combined with an unexpected (at least on a first visit) gift. Definitely a good idea from the sensory marketing perspective. I must confess that I am a regular visitor, and I can't help but worry that my exposure to the high-energy sweet food scent encourages me to consume a cookie that I might otherwise not have eaten.

My grandfather, who had a grocer's shop in the north of England, would sprinkle quality coffee beans behind the counter (see Figure 2.5). When a customer came into the shop he would crush the beans underfoot, releasing the coffee aroma that would hopefully nudge them into buying some of the real stuff. So, given what I have just told you, you will understand why it does not come as any surprise to me to see food stores now releasing the enticing scent of their

Figure 2.5. My grandfather's shop in Idle, Bradford, where olfactory sensory marketing was being used intuitively half a century ago.

products in order to try to lure customers in. But is all this scent-based marketing actually provoking our desire to eat in many situations where we might not otherwise have thought about consumption?

We really do need to be more aware of the consequences of the looming commercialization of food-based scent and flavour marketing. Those exposed to food aromas exhibit an increased appetite not only for any food that happens to be associated with that specific aroma but also for other foods and beverages that are similar in terms of their macronutrient profile. That is, exposure to one sweet high-energy food leads to our appetite for other foods with a similar aroma increasing. Have you noticed how food chains tend to locate their stores in those positions in a mall that ensure the optimal dispersion of their signature smell, and often combine this strategy with the use of the least effective extractor fan permitted, so that more of the aroma hits the consumer's nostrils.[7]

Let me leave you for now, though, with an even scarier thought. Have you ever considered what is going to happen when food runs out as a result of global warming, overfishing and food blights? Well, artists Miriam Simun and Miriam Songster have imagined how three foods that are currently threatened, namely chocolate, cod and peanut butter, might be enjoyed in the future. To illustrate their take on the future, they drove *The Ghost Food Truck* from Philadelphia to

New York in 2013. Those who visited this most unusual piece were given a mask to wear that allowed them to smell what it was that they were supposed to be eating. According to one commentator: 'When you get your sample, it comes with something that looks a bit like a medical breathing tube. Fit it around your face, and it holds a small bulb soaked in synthetic chocolate, cod, or peanut butter scent right next to your nose. Once you finish eating [vegetable protein and algae], the attendant pops off the bulb and takes it away, cleaning the frame for the next guest.'[8] Should this dystopian view of the future one day come to pass, then F. T. Marinetti's conception of plates of food that you would be allowed to smell but not eat from more than a century ago (see the final chapter for more on the Futurists) may turn out to be closer to the truth than any of us realized. In summary, then, there really is no escaping the value that understanding the nose brings to the way we connect to our plates.

In the next chapter, we are going to explore sight. We will take a closer look at the phenomenal rise of gastroporn and the recent emergence of *mukbang* in the Far East. What's that, I hear you ask. Well, I'm afraid you will just have to wait to find out . . .

3. Sight

Your brain is your body's most blood-thirsty organ, utilizing around 25% of total blood flow (or energy) – despite the fact that it accounts for only 2% of body mass. Given that our brains have evolved to find food, it should perhaps come as little surprise to discover that some of the largest increases in cerebral blood flow occur when a hungry brain is exposed to images of desirable foods. Adding delicious food aromas makes this effect even more pronounced. Within little more than the blink of an eye, our brains make a judgement call about how much we like the foods we see and how nutritious they might be. And so you might be starting to get the idea behind gastroporn.

No doubt we have all heard our tummies gurgling when we contemplate a tasty meal. Viewing food porn can induce salivation, not to mention the release of digestive juices as the gut prepares for what is about to come. Simply reading about delicious food can have much the same effect. In terms of the brain's response to images of palatable or highly desirable foods (food porn, in other words), research shows widespread activation of a whole host (or network) of brain areas, including the taste (i.e., gustatory) and reward areas (the insula/operculum and orbitofrontal cortex respectively). The magnitude of this increase in neural activity, not to mention the enhanced connectivity between different brain areas, typically depends on how hungry the viewer is, whether they are currently dieting (i.e., whether they are a restrained eater or not) and whether they are obese. (The latter, for instance, tend to show a more pronounced brain response to food images even when full.)

Apicius, the first-century Roman gourmand and author, is credited with coining the aphorism: 'The first taste is always with the eyes.' Nowadays, the visual appearance of a dish is just as important as, if not more important than, the taste/flavour itself. We are bombarded by food images everywhere, from adverts through to social media and TV cookery shows. There is simply no getting away from them. Unfortunately, though, the foods that tend to look best (or rather, that our brains are most attracted to) are generally not the healthiest. Quite the reverse, in fact, as we will see later.

We may all face being led into less healthy food behaviours by all of the highly desirable images of foods that increasingly surround us nowadays. In 2015, just as in the year before, food was the second most searched-for category on the internet (after pornography). Thus the blame, if any, doesn't reside solely with the marketers, food companies and chefs; a growing number of us are actively seeking out images of food – a kind of 'digital foraging', if you will. How long, I wonder, before food takes the top slot?

Can you taste the colour?

What we taste is profoundly influenced by what we see. Similarly, our perception of aroma and flavour are also affected by both the hue (i.e., red, yellow, green, etc.) and the intensity, or saturation, of the colour of the food and drink we consume. Change the colour of wine, for instance, and people's expectations – and hence their tasting experience – can be radically altered. Sometimes even experts can be fooled into thinking that they can smell the red wine aromas when given a glass of what is actually white wine that has just been coloured artificially to give it a dark red appearance!

At various points in history, scientists (some of them really rather eminent, like one of the godfathers of psychology, Hermann Ludwig Ferdinand von Helmholtz – anyone with a name that long should probably not be messed with!) have confidently asserted that there is absolutely *no* association between colour and taste. At the other extreme, there are some artists out there today who are inviting the public to 'taste the colour'.[1] I have to say that I don't think either side has got it quite right. Colours are very definitely linked with tastes, and yet I do not believe that you can create a taste out of nowhere, simply by showing the appropriate colour.

Consider the amuse bouche created by the London-based chef

Jozef Youssef as part of Kitchen Theory's 'Synaesthesia' dining events, which developed out of the latest research findings from my Crossmodal Research Laboratory. We have spent the last few years researching what tastes people around the world associate with different colours, and looking at the colours that people naturally/spontaneously associate with the four most frequently mentioned basic tastes. The results fed directly into the design of the chef's dish. Had you been a guest at the restaurant, you would have had four spoons of espherified colourful taste placed randomly down in front of you – one red, one white, one green and one browny-black. Once everyone has their four spoons, they are informed that the chef recommends starting with the salty, then the bitter spoon, next the sour, and ending on a sweet note – leaving you and the other diners a little perturbed as to which order exactly you should taste the spoons in. The idea is that the diners arrange their own spoons from left to right in front of them in the order: salty, bitter, sour and sweet. Having arranged their own spoons, the diners normally start to look around and compare notes with each other. In the restaurant, or online, we get somewhere around 75% of people ordering the spoons in the way that the chef (and the gastrophysicist) intended. So on the basis of such results, I would say that tastes are very definitely associated with specific colours.

Colour can be used to modify people's perception of a taste that is already present in the mouth. I can, for example, make food or drink taste sweeter by adding a pinkish-red colour, but, as yet, I haven't come across a way to do that while serving someone a glass of water. Turning water into wine – now that's still a step too far, even for the gastrophysicist at the top of their game (though, if you remember, in the previous chapter, that was pretty much what the inventors of The Right Cup were promising to do)! Nevertheless, a food or beverage company might be able to increase perceived sweetness by up to 10% by getting the colour of their product, or the packaging in which it comes, just right. Every little counts, as they say.

Some people want to know whether the effects of adding colour are like adding sugar. Surely psychologically induced sweetness must taste different from the chemically induced kind? Well, the results of side-by-side tests show that people will sometimes rate an appropriately coloured drink (imagine a pinkish-red drink) as sweeter than an inappropriately coloured (say, green) comparison drink. Such results can be obtained even if the latter drink has as much as 10% more added sugar. In other words, psychologically induced taste enhancement is indeed indistinguishable from the real thing, at least sometimes. Sweetness without the calories – now who wouldn't want that?[2]

Our responses to colour in food and drink aren't fixed but change over time. For instance, a few decades ago, marketers and cultural commentators were telling anyone who'd listen that blue foods would *never* sell.[3] Roll the clocks forward a few decades, though, and we now have cool blue Gatorade, Slush Puppy and the London Gin Company all successfully promoting blue drinks. A Spanish company even launched a blue wine in 2016.* Given the rarity of this colour in nature, it is normally introduced solely as a marketing ploy to capture the attention of consumers by standing out on the store shelf. The problem often comes later, though, when people actually get to taste that eye-catching drink. Seeing a transparent blue colour will put a certain idea in the consumer's mind about the likely taste. And if the expectation doesn't match up with the reality, the manufacturer may well have a problem on its hands.

In fact, you would probably be surprised to learn how many companies have come looking for help over the years, when their consumer panels and focus groups tell them that their brands taste different, even though all that has changed is the colour of the product or pack. For example, a mouthwash manufacturer told me that their orange variant didn't taste as astringent to people as their regular blue variety, despite the formulation of the active ingredients staying the

---

* This is such a bad idea that I am pretty sure it won't still be around by the time you read this. Maybe those old marketers weren't completely wrong after all. The drink is supposedly targeted at millennials who, so we are told, like brightly coloured alcopops, etc. But I am really not sure that they will want to be seen drinking blue wine. Ironically, the Italian Futurists used to serve their guests blue wine to shock them. And to think that someone now believes it will sell!

same. It makes no sense until you learn something about the rules of multisensory integration governing how the brain combines the senses. Here, I am thinking of 'sensory dominance' – where the brain uses one sense to infer what is going on in the others.

The impact of colour depends on the food. In the context of meat and fish, blue causes a distinctively aversive response. In one of my favourite 'evil' experiments – the sort of thing that ethics panels were surely brought in to put a stop to – a marketer by the name of Wheatley served a dinner of steak, chips and peas to a group of friends. At the start of the meal, the only thing that might have struck anyone as odd was how dim the lighting was. This manipulation was designed to help hide the food's true colour. For when the lights were turned up part-way through the meal, Wheatley's guests suddenly realized that they had been tucking into a steak that was blue, chips that were green and peas that were bright red. A number of them apparently started to feel decidedly ill, with several heading straight for the bathroom![4]

Can you taste the shape?

This is another of those questions to which the intuitive answer has to be: 'But of course not.' And yet, over the last decade I have had lots of fun at food and science festivals around the world, giving people different foods to taste while asking them to tell me whether some aspect of the tasting experience is more 'bouba' or 'kiki'? If you don't have the faintest idea what I am talking about, don't worry. Just take a look at the shapes shown at either end of the scale in Figure 3.1, and ask yourself which one would you give each of those made-up names. Most people say that 'kiki' must be the angular shape while 'bouba' is obviously the rounded blob.

Next, I would like you to imagine the taste of dark chocolate, Cheddar cheese or sparkling water. Place an imaginary mark on the scale that, in some sense, 'matches' the sensory properties of the experience of each of those foods. Sounds silly, right? But just try it. Do the same thing for milk chocolate, a nice ripe piece of Brie or still

Figure 3.1. Do tastes, aromas and flavours have shapes? This figure illustrates the simple shape symbolism scale that we often use in our research. The crayon indicates the mid-point of the scale.

water – the choice is yours. My gastrophysics research indicates that you will probably have marked those last three more towards the left (bouba) end of the scale, whereas the other three will be more towards the right.

What's fascinating about this kind of research is just how consistent people's answers are (given that there are no objectively correct answers to these questions). Most people will place carbonated, bitter, salty and sour-tasting foods and drinks towards the angular end of the scale. Sweet and creamy sensations, by contrast, are nearly always paired with rounder shapes. In other words, we all seem to show a generalized tendency to want to match specific shapes (or contour) with particular tastes, aromas, flavours and even food textures.

Why is it that people associate shapes with tastes at all? One evolutionary account for this is that angular shapes, bitterness and carbonation are all somehow linked with danger or threat. An angular shape could be a weapon, bitterness might indicate poison, while sourness and carbonation would, once upon a time at least, have been an avoidance cue signalling overripe or spoiled foods. By contrast, sweet and round both have positive connotations. So maybe we just choose to put together those stimuli that we feel the same way about, even if we have never experienced them together before. Alternatively, there may be something in the environment – some correlation between shape and taste properties – that we (or rather, our brains) are picking up on. I sometimes wonder whether the acidity in cheese might be correlated with its texture. Could it be that harder cheeses (i.e., those that are more likely to retain their angular form on cutting) are, on average, more acidic than soft and runny cheeses? Does dark chocolate tend to break into more angular shapes than milk chocolate?

Next time you are in the supermarket, take a look at the beer or bottled water shelves. The majority of beer and carbonated water brand logos are angular, not round. There are exceptions, of course, but it is remarkable how often you see a red star or triangle on the front of a bottle or can – the many red stars that adorn the San Pellegrino sparkling water bottle, or the prominent red star on the Heineken label, for example. See how the food and beverage industry are communicating with you in ways that are functionally subliminal? But apart from the marketing angle, just how important is knowing about shape symbolism? Well, I have always been most interested in what happens to taste when you start manipulating the shape of food itself. Back in 2013, Cadbury decided to update the shape of the iconic Dairy Milk chocolate bar by rounding off the corners, reducing the weight of the chocolate by a few grams in the process. Consumers wrote in and called in droves to complain. They were convinced that the company had changed the formula – that their favourite chocolate now tasted sweeter and creamier than before. But a spokesperson from Mondelez International (the American owner of Cadbury) stated: 'We have been very clear and consistent that we have not changed the recipe of the much-loved Cadbury Dairy Milk, although it's certainly true that we changed the chunk last year from the old, angular shape to one that's curved.'[5] So you can imagine how companies could (if sugar weren't so cheap) perhaps reduce the sugar content while making the shape rounder. The product itself would be a little less unhealthy, and it's just possible that the taste wouldn't change in the mind of the consumer either. It is a challenge, obviously, to get companies to reformulate their products effectively, but it can be done.

What is true of chocolate turns out to apply to other foods as well. When you serve food, be it a beetroot jelly or a chocolate confection, in a rounder shape, people often rate it as tasting sweeter than when exactly the same food is presented in a more angular form instead.

In fact, even categorizing angular (rather than round) shapes on a piece of paper just before eating influenced the rated sharpness of a piece of Cheddar cheese in one North American study.[6] Similarly, varying the shape made by the chocolate sprinkled on top of a caffè latte can be all it takes to modify the drinker's taste *expectations*. We have been working with baristas in Australia to show that people expect a caffè latte with a star shape sprinkled on top to taste a little more bitter than a latte with a rounded, bouba-esque shape sprinkled on top.[7] Though whether that change in expectations is sufficient to alter the perceived taste depends on how close the taste of the drink is to what was predicted by the customer.

When the taste is pretty much as expected, the shape may well influence what people say about the taste. However, if there is too much of a divergence, the brain seems to discount the shape cue altogether. That, at least, is our current hypothesis in the lab. And, of course, even when the shape does change the taste, whether that results in increased liking or not is going to depend on the taste of the food itself, and on the taster. While sweetness is normally liked, we conducted one study in a Scottish hotel restaurant where an overly sweet dessert was actually liked less when it was served on a 'sweeter'-looking plate.[8]

Can you taste the plate?

Most people will simply shake their heads when you ask them whether changing the colour of a plate would influence the taste of the food on it. However, in research conducted together with the Alicía Foundation in Spain, exactly the same frozen strawberry mousse was rated as tasting 10% sweeter and 15% more flavourful, and was liked significantly more, when eaten from a white plate rather than from a black plate instead (see Figure 3.2). Remarkably, follow-up research by scientists working in Greenland obtained even more striking results by varying both the colour and the shape of the plate. Strange though it may seem, round plates are just 'sweeter' than angular plates!

In other research, we have been able to modify the taste of everything from a vended cup of hot chocolate to a caffè latte by varying

Figure 3.2. White and black plates with red frozen strawberry dessert. The funny thing is that people rated the dessert from the white plate as tasting significantly sweeter and more flavourful than exactly the same food served on the black plate. Such results certainly make me wonder about the sense of using different coloured charger plates, as in Denis Martin's namesake restaurant in Vevey, Switzerland (remember the cow?).

the colour of the cup. Hot chocolate tastes more, well, chocolatey, and is liked significantly more, when served from an orange plastic cup (rather than a white one). Meanwhile, caffè latte is judged more intense while the sweetness is somewhat suppressed when served in a white porcelain rather than a clear glass mug.

Intriguingly, gastrophysics research also shows that enhancing the visual contrast on the plate can lead to a substantial increase in food and liquid intake in those suffering from advanced Alzheimer's disease. In one study conducted at a long-term-care facility in the US, for instance, switching to high-contrast coloured plates and glasses led to a 25% increase in food consumption and liquid intake going up by as much as 84%! The results of another hospital study were equally dramatic: average consumption amongst the older and more vulnerable patients, including those suffering from dementia, went up by 30%, just by changing the plate colour! So impressed were they with the results in this case that the hospital went on to replace their standard-issue white plates with blue crockery.

But hold on a minute, I hear some of you say. There is something that doesn't quite make sense here. What about the 'blue plate special'? As the wonderfully named Bunny Crumpacker put it in her book *The Sex Life of Food*: 'the term blue plate special became popular during the Great Depression because restaurant owners found that diners were satisfied with smaller portions of food if it was served on blue plates.'[9] So how can it be that blue plates reduced consumption back in the 1920s but significantly increase how much patients eat today? One explanation might be that the majority of the food served in hospitals tends to be bland in both taste and colour, so, it simply fails to stand out against a white plate. By contrast (boom-boom), served on a blue plate it is suddenly much easier to see what one is eating. This is precisely why, at home, I used to like serving my signature Thai green chicken curry (greenish-white in colour) with white rice on a black plate: the visual contrast was much more striking. Or as one article I read put it, serving steak on a white plate is fine, but porridge never should be. Though, sadly, the delightful Mrs Spence isn't of the same opinion. She thought the angular black plates were a little too batchelor-ish, so out they went. So much for gastrophysics!

Even to a non-gastrophysicist, presenting hospital food on a red plate or tray must seem intuitively wrong. There is a justification for this approach, of course. It is supposed to help the relevant healthcare professionals more easily identify those patients needing special nutritional attention. I suspect that it is a bad idea, though. Why? Well, because the colour red tends to elicit avoidance motivation. What this means in practice is that people eat significantly less when served food on a red plate than when offered exactly the same food from plateware of another colour. And the effects aren't small either.

In one study, people consumed almost twice as many pretzels (under lab conditions) when they ate from a white plate than from a red plate. There seems no good reason to imagine that serving food on a red tray, rather than from a tray of a different colour, wouldn't trigger exactly the same kind of avoidance motivation. So red plates and trays might be recommended for anyone who wishes to lose weight, but this is simply not the situation that most hospital patients find themselves in.

So, while you can't literally taste the plateware on which food is served, its colour (not to mention its size) is likely to modify your behaviour. Perhaps it might bias you to eat more (or less) than you otherwise might have done. What is more, it is also likely to influence your experience of whatever is served, possibly making it seem more delicious, sweeter or more flavourful.

The effect of background colour on taste and flavour perception causes all manner of problems for the food and beverage industry, as a number of companies have found to their cost. Simply altering the can colour – say, adding a little more yellow to the side of 7Up (as was done in the 1950s), or launching a white Christmas Coke can (as happened in 2011) – can change what consumers think about the taste. Of course, modifying the colour of the packaging shouldn't affect the taste of the contents, especially for a brand that people are already familiar with. And yet the evidence suggests that it most certainly does! The phenomenon provides an example of visual dominance (of the packaging colour) – which may, in fact, be the only product-relevant sensory cue that the consumer has to go on before they start drinking directly from the can.

Now, take a look at the plate shown in Figure 3.3. What do you think serving dinner from such an odd bit of plateware will do to the experience? Given the emerging gastrophysics research, no one should be surprised by the explosion of interest in enhancing people's enjoyment of food by breaking away from the rigid tyranny of the large, round, white plate (what some refer to as the American plate). No matter whether you are a diner or a chef, you can benefit from paying a little more attention to the latest gastrophysics research and to the plateware itself, not just what is on it.

It used to be that only the world's top restaurants served each of the dishes on their multi-course tasting menus from dedicated plateware but this is now slowly percolating down through chain restaurants all the way to the adventurous home chef. Are we reaching the point where the fact that a dish looks good (or photographs well) is becoming more important to its success than how it tastes? This growing trend towards gastroporn has been parodied by Canadian chef Carolyn Flynn, aka 'Jacques La Merde'. La Merde has an Instagram site with over 100,000 followers which shows food picked up from the

Figure 3.3. A fabulous piece of plateware, as used at the Montbar and Tickets restaurants in Barcelona. Serving food on one of these plates may well change what people say about the taste. Ice cream, for instance, might taste that little bit sweeter when served on this than if the same product were to be served on an angular black slate.

convenience store or fast-food restaurant – i.e., junk food like crumbled Oreos and Doritos – plated up so as to look like it came from a top Michelin-starred restaurant.[10]

Ideally, of course, you want both. Doesn't everyone want food that not only looks beautiful but tastes great too? There are a couple of important challenges here for the gastrophysicist. One is to develop robust experimental tests to assess the impact of changes in plating on people's perception (and how much they would be willing to pay). But the gastrophysicist can also provide theoretical insights into the aesthetics of plating (i.e., not just analysing what the chefs have done to see what people like but also making predictions/recommendations about the sorts of configurations that the diner ought to like, based on what we already know about the brain). Taken together, these will hopefully enable those who are preparing really great-tasting food to come up with plating arrangements that have maximum visual appeal as well. Delivering on that promise is easier when the chef combines their skills in the kitchen with the latest gastrophysics testing techniques and insights.

When did 'food porn' first appear and where is it heading?

People have been preparing beautiful-looking foods for feasts and cele-
brations for centuries. (And, of course, artists have for centuries
captured them in all those still life paintings too.) However, for anything other than an extravagant feast, the likelihood is that meals in
the past would have been served without any real concern for how
they looked. That they tasted good, or even just that they provided
some sustenance, was all that mattered. Now, this was true even of
famous French chefs, as highlighted by the following quote from
Sebastian Lepinoy, Executive Chef at L'Atelier de Joel Robuchon,
describing the state of affairs before the emergence of nouvelle cuisine:
'French presentation was virtually non-existent. If you ordered a coq
au vin at a restaurant, it would be served just as if you had made it at
home. The dishes were what they were. Presentation was very basic.'[11]

Everything changed, though, when East met West in the kitchens
of the French cookery schools in the 1960s. It was this meeting of
culinary minds that led to nouvelle cuisine and, with it, gastroporn.
The latter term dates back to a witty review written in 1977, describing Paul Bocuse's *French Cookery* cookbook as a 'costly ($20.00)
exercise in gastro-porn'.[12] The term stuck, and has now made its way
into the *Collins English Dictionary*, which defines it as 'the representa-
tion of food in a highly sensual manner'. Others prefer the term 'food
porn'. Make no mistake, though, they are all talking about the same
thing.

These days, more and more chefs are becoming concerned
(obsessed, even) by how their food photographs. And not only for the
glorious full-colour, full-page pictures that will adorn the pages of
their next glossy cookbook. As one restaurant consultant put it: 'I'm
sure some restaurants are preparing food now that is going to look
good on Instagram.'[13] Certainly, chefs who serve something visually
stunning and plated beautifully, or else plated on or in something
most unusual, like a brick, trowel or flat cap, can give their dishes
some real eye-appeal, and hence hopefully help grow their digital
presence.

Figure 3.4. Foodography – it's all about getting the perfect shot. The curved plate and smartphone stand make taking a great photo a cinch.

Some chefs have been struggling with how to deal with the growing trend for diners to photograph their food and share it on social media. Their much publicized responses include everything from limiting their diners' opportunities to photograph the food during the meal through to banning photography inside the establishments they preside over completely. The latter approach seems doomed to failure, for you just can't fight the tides of change. Best to get with the program and figure out how to adapt your food offering for the growing number of experience-hungry millennials out there who want to share their every waking moment with their social networks. It would, however, seem as though the chefs have now, mostly, embraced the trend, acknowledging that it is all part of 'the experience'. As Alain Ducasse, chef at London's three-Michelin-starred Dorchester Hotel says: 'Cuisine is a feast for the eyes, and I understand that our guests wish to share these instants of emotion through social media.'[14]

When thinking of a more constructive approach to resolving this looming technology-driven crisis in high-end cuisine, just consider those cutting-edge restaurants that have now started to incorporate specially shaped plates into their service in order to provide the perfect backdrop for the food in their diners' photos. Others, like Foodography, at Catit, a restaurant in Tel Aviv, Israel, offer their diners camera stands at the table, or else serve their food on plates that spin 360 degrees, thus promising their customers the perfect shot (see Figure 3.4).

Spinning plates might well strike you as a bit over the top. But then again, haven't we all subtly, almost unconsciously, rotated the plate once or twice after a waiter has placed a dish in front of us in a restaurant? Have a look at the two plates of food shown in Figure 3.5 and tell me which one you prefer. The left one, right? Yet all that differs is the orientation! Our latest online research shows that people's preference for a plate of food can vary quite dramatically simply as a function of the orientation in which it is presented. In our online experiments, for instance, we find that people are willing to pay significantly more for exactly the same food when shown in one orientation as compared to another. Now that you know this, don't you want to optimize the orientation of the plate every time you serve food? If such a simple step can genuinely enhance the

Figure 3.5. The same plate of food served two ways. Several thousand people have been spinning this particular plate, the signature dish from Brazilian chef Albert Landgraf, into the perfect orientation (over the internet, in a massive citizen science experiment). Based on the results, we now know that the majority of people prefer the plate shown on the left. Ideally, the onions should be plated so that their tips point 3.4 degrees past 12 o'clock. The chef had decided to plate it at 12. So the chef intuitively got it more or less spot on. No need for the gastrophysicist, other than for confirmation, in this case!

perceived value and enjoyment of the food you serve, no matter who you are and no matter whom you are serving, you'd be crazy not to.

Beyond that, there is a sense in which the visual appeal of the meal has become an end in and of itself, with a growing number of people taking images of what they eat. Certainly, these days the press is full of tips on how to make your food photography more visually attractive, offering, as one newspaper did recently, to help you 'turn your dull food images into Instagram food porn in 12 simple steps'.[15] Researchers and food companies have, in recent years, begun to establish which tricks and techniques work best in terms of increasing the eye-appeal of a dish, including, for instance, showing food, and especially protein, in motion (even if it is just implied motion) to attract the viewer's attention and convey notions of freshness. The gastrophysicist knows only too well just how important this is: a beautifully plated dish is likely to taste better than it might if the elements were randomly plonked on the plate. At the same time, I and many of my colleagues who work as chefs worry that this increase in eye-appeal sometimes comes at the cost of the optimization of the actual flavour of the dish.[16]

Have you heard of 'yolk-porn'?

What do you get if you show protein (e.g., oozing egg yolk) in motion? Answer: yolk-porn. Seriously! This is the all-new trend in addictive food images (see Figure 3.6). I came across an example recently in a London underground station. There were a whole host of video advertising screens along the wall as I ascended the escalator. All I could see, out of the corner of my eye, was a steaming slice of lasagne being lifted slowly from a dish, dripping with hot melted cheese, on screen after screen. As the marketers know only too well, such 'protein in motion' shots are very attention-grabbing; our eyes (or rather our brains) find them almost irresistible. Images of food (or more specifically, energy-dense foods) capture our visual awareness, as does anything that moves. 'Protein in motion' is therefore precisely the kind of energetic food stimulus that our brains have evolved to detect, track and concentrate on visually.

Figure 3.6. Just a toast soldier being harmlessly dipped into an egg yolk, do you think, or something rather more pernicious?

The British retailer Marks & Spencer have acquired something of a reputation for their food porn with much of their advertising over the last decade or so including highly stylized and gorgeously presented visual imagery. Look closely at their ads and one finds plenty of protein in motion (both implied and real). Their most famous ad, from 2005, was for a chocolate pudding shown with an extravagant melting centre. A sultry voiceover came out with the iconic – though subsequently much parodied – line: 'This is not just chocolate pudding, this is a Marks & Spencer chocolate pudding.' But do you have any idea what this one ad did to sales? They skyrocketed by around 3500%!

What was especially noticeable about M&S's 2014 campaign (which paid homage to the original gooey pudding) was that *all* of the food was shown in motion. In fact, one of the most widely commented on images was of a Scotch egg being sliced in half, with the yolk oozing out slowly. According to one company executive, the ad showcased 'the sensual and surprising aspects of food – like its textures and movement – in a modern, stylish and precise format'. And M&S certainly aren't the only ones to have adopted this approach. An informal analysis of the food commercials shown in the highly valued advertising slots during the US Super Bowl revealed that in the years 2012–14 two-thirds showed food in motion. The attraction that we all feel towards images of food in motion might also help to explain the viral explosion of interest in one particular video of a melting chocolate dessert that swept the internet recently, described by a number of journalists as simply 'hypnotic'.

One other reason to show food in motion (real or implied) is that it looks more desirable, in part because it is perceived to be fresher. For instance, food psychology and marketing researcher Brian Wansink and his colleagues at Cornell University conducted research showing that we rate a picture of a glass of orange juice as looking significantly more appealing when juice can be seen being poured into the glass than when the image is of a glass that has already been filled. Both are static images but one *implies* motion. That was enough to increase the product's appeal. (For those of you at home, who may not be able to guarantee that your food moves, another strategy is simply to leave the leaves and/or stems on fruit and vegetables, to help cue freshness.)

Mukbang

Now let me tell you, dear reader, about one of the strangest trends relating to food porn that I have come across in recent years; it's called *mukbang*.[17] A growing number of South Koreans are using their mobile phones and laptops to watch other people consuming and talking about eating food online. Millions of viewers engage in this voyeuristic habit, which first appeared back in 2011, every day. Interestingly, the 'porn' stars, or 'broadcast jockeys' (BJs) as they prefer to be known, are not top chefs, TV personalities or restaurateurs but rather just regular (albeit generally photogenic) 'online eaters' (see Figure 3.7). One can think of this as yet another example of food in motion; it's just that the

Figure 3.7. *Mukbang* – it roughly translates as 'food porn'. The live-streaming of people eating food started in South Korea in 2011 and now attracts millions of viewers.

person interacting with the food happens to be more visible than in many examples of dynamic food advertising in the West, where all you see is the food moving – just think of those M&S ads. I also get the sense, though, that some of those who, for whatever reason, often end up eating alone are tuning in for a dose of *mukbang* at mealtimes to get some virtual company (see the 'Social Dining' chapter).

It would be interesting to see whether those who eat while tuning in to one of these increasingly popular BJs end up consuming more than they would were they really eating alone (i.e., without any virtual dinner guests). One might also wonder if *mukbang* is as distracting as regular television, which has been shown to dramatically increase the amount consumed. If so, one might expect not only that the viewer's immediate intake of food will increase but also that the amount of time that passes before they get hungry again ought to be reduced.

With *mukbang*, viewers can imagine themselves having a meal with whoever they see on the screen. However, research shows that food imagery is most visually appealing when the viewer's brain finds it easy to simulate the act of eating, i.e., when the food is seen from a first-person perspective. This is rated more highly than viewing food from a third-person view (as is typically the case with *mukbang*). Marketers, at least the smarter ones, know only too well that we will rate what we see in food advertisements more highly if it's easier for our brains to mentally simulate the act of eating that which we see. So, for example, imagine a packet of soup. A bowl of soup is shown on the front of the packaging. Adding a spoon approaching the bowl from the right will result in people being around 15% more willing to purchase the product than if the spoon approaches from the left instead. Why so? Well, it's because the majority of us are right-handed, and so we normally see ourselves holding a spoon in our right hand. Simply showing what looks like a right-handed person's spoon approaching the soup makes it easier for our brains to imagine eating. Now, for all those lefties out there saying, 'What about me?' – it may not be too long before the food ads on your mobile device might be reversed to show the left-handed perspective. The idea is that this will help maximize the adverts' appeal to you (assuming, that is, that your technology can figure out your handedness implicitly).

How worried should we be by the rise of food porn?

Is there anything to be worried about here, or is it all just a storm in a teacup (or should that be Twitter-feed)? Why shouldn't people indulge their desire to view all those delectable gastroporn images. Surely there is no harm done? After all, food images don't contain any calories, do they? Well, it turns out that there are a number of problems that gastrophysicists have proven and that I think we should be concerned about:

1. *Food porn increases hunger.* One thing that we know for certain is that viewing images of desirable foods provokes appetite. For example, in one study, simply watching a 7-minute restaurant review showing pancakes, waffles, hamburgers, eggs, etc. led to increased hunger ratings not only in those participants who hadn't eaten for a while but also in those who had just polished off a meal.[18] As Italian researchers put it: 'Eating is not only triggered by hunger but also by the sight of foods. Viewing appetizing foods alone can induce food craving and eating.'[19]

2. *Food porn promotes unhealthy food.* Some of Nigella Lawson's delicious-looking cakes, the ones you see her making on TV, contain in excess of 7000 calories. In fact, many of the recipes that top chefs make on television cookery shows are incredibly calorific or unhealthy (despite chefs' pronouncements about healthy eating in the press!). Indeed, those who have systematically analysed the TV chefs' recipes find that they tend to be much higher in fat, saturated fat and sodium than is recommended by the World Health Organization's nutritional guidelines.[20] This is not only a problem for those viewers who go on to make the foods that they see their idols making. (Surprisingly few people actually do this; maybe it is not such a problem after all. According to a recent survey of 2000 foodies, fewer than half had ever cooked even one of the dishes that they had seen prepared on food shows.) Rather, the bigger concern here is that the foods we see being made, and the food portions we see being served on TV may set implicit norms for what we consider it appropriate to eat at home or in a restaurant.

3. *The more food porn you view, the higher your Body Mass Index (BMI).* While the link is only correlational, not causal, the fact that people who watch more food TV have a higher BMI might nevertheless cause you to raise your eyebrows. They might, of course, be watching more television generally, not just food programmes – after all, the term 'couch potato' has been around for longer than the term 'food porn'. The key question, though, from the gastrophysics perspective, is whether those who watch more food television have a higher BMI than those who view an equivalent amount of non-food TV. It would certainly seem likely to be the case, given all the evidence out there showing that food advertising biases subsequent consumption, especially in kids.

4. *Food porn drains mental resources.* Whenever we view images of food – on the side of product packaging, in cookery books,

or increasingly on TV shows or social media sites like Instagram's 'Art of Plating' – our brains can't help but engage in a spot of embodied mental simulation. That is, they simulate what it would be like to eat the food. At some level, it is almost as if our brains can't distinguish between images of food and real meals. We therefore need to expend some mental resources, silly though it may sound, to resist all of these *virtual* temptations. So what happens when we are subsequently faced with an actual food choice? Imagine yourself watching a TV cookery show (such as on The Food Network) and then arriving at a train station; the smell of coffee wafting through the air leads you by the nose into buying a cup. While at the counter, you see the sugary snack bars and fruit laid out in front of you. Should you go for a bar of chocolate, or pick a healthy banana instead? In one laboratory study, participants who had been shown appealing food images tended to make worse (i.e., more impulsive) food choices afterwards than those who had been pre-exposed to a smaller number of food images. All of this (increased and increasing) exposure to desirable food images results in involuntary embodied mental simulations. That is, our brains imagine what it would be like to consume the foods that we see, even if those foods are only on the TV or our smartphones, and we (or our brains) then have to try to resist the temptation to eat. One recent study, conducted in three snack shops located in train stations, investigated whether people could be nudged towards healthier food choices simply by moving the fruit closer to the till than the snacks – the reverse normally being the case. The 'nudge' worked in the sense that people were indeed more likely to buy a piece of fruit or a muesli bar. Unfortunately, though, people continued to purchase snacks like crisps, cookies and chocolate *as well*. In other words, an intervention that had been designed to *cut* people's consumption actually resulted in their consuming *more* calories (assuming that what was purchased was consumed)![21]

Can I engage in a spot of gastroporn at home?

The good news is that labs like mine have been researching ways to turn this visual attraction to food into a positive, so that we can learn to make healthier meals at home. Any one of us can make the food we serve more attractive by plating for the eye. In one of the studies conducted here in Oxford, for example, we served a salad in a college dining room to 160 diners either just as a regular tossed salad, or with the elements arranged to look a little like one of Kandinsky's paintings. The results showed that the diners were willing to pay more than twice as much for exactly the same food when it looked more visually attractive (see Figure 3.8). The complete Kandinsky salad might look daunting to try to create at home (given that it has more than thirty elements). But even a typical UK at-home meal of steak and chips and a garden salad (with only three ingredients) can be made significantly more desirable by paying a little more attention to the presentation.[22]

Figure 3.8. Same food, alternative plating. Diners were willing to pay twice as much for the food shown on the left despite it having exactly the same ingredients as the less artfully arranged salad on the right. Once you realize how much difference even simple changes to the eye-appeal of a dish can make, it seems crazy not to take a few tips from the gastrophysicists and serve something that is optimally visually attractive. What have you got to lose?

Another trick here is to use the visual appearance of food to help yourself eat less. One idea is to serve it off a smaller plate (thus making it look like there is more food). One should avoid serving food on a plate that has a wider rim too. Even looking at the picture of the bowl on the front of a packet of breakfast cereal can influence how much we think it appropriate to serve ourselves (i.e., it acts as a serving/consumption norm). Show the cereal in a bowl with a rim and it will look like a smaller portion than if shown in a rimless bowl. In some of our latest research, we have been able to demonstrate that showing a given amount of cereal in a bowl with no rim versus a wide rim really can make a difference. According to researchers from Oxford and Cambridge, reducing the size of the plate or bowl we eat from reduces average calorie intake by roughly 10% (or 160 calories).[23]

Imagined consumption

Imagine yourself eating lots of M&Ms. Do you think that this could affect how much you would end up eating if offered a bowl of the brightly coloured sweets afterwards? Research shows that you would eat significantly fewer of them if they were subsequently offered to you. Thus, the *imagined* consumption of food can reduce actual consumption.[24] Those of you on a diet, take note! Simply by repeatedly imagining the act of consumption, you are likely to then eat a little less. This habituation effect is, however, food specific. Getting people to imagine eating a different food, like cheese, doesn't, I'm afraid, suppress the desire to eat, say, chocolate.

Similarly, getting people to remember the last meal that they ate also suppresses subsequent snacking behaviour. It is, of course, theoretically possible that just seeing a food once could result in the viewer imagining the act of eating it, and hence give rise to reduced consumption due to habituation. However, in practice this does not seem to be the case, presumably because no one in their right mind would spend their time imagining eating the same food sixty times in a row! (Or just thirty times in one of the other studies.) Instead, the exposure to palatable food images (in the absence of instructions to repeatedly imagine eating the seen food) generally, though not always, leads to increased consumption of whatever food is made available thereafter. Though I do wonder whether, in much the same way that watching pornography may reduce a viewer's sexual activity, there might be situations under which the same could be true for at least some of those individuals who watch large amounts of food porn – i.e., they use it to in some sense control their cravings for the real stuff.

Ugly fruit

Those who are really serious about losing some weight could do worse than to make their food look as ugly as possible.* At the opposite extreme we have enhanced 3D virtual reality food blogs (like Perception Fixe, by Matheus DePaula-Santos of Myo Studios) whose aim is to make food look as desirable as possible. According to one journalist: 'Myo Studios is banking on the notion that providing an enhanced visual experience through virtual reality will markedly up its food blog's ante. Users will be able to "sit down in front of a steak from some restaurant, even though there's no reservation for three months." DePaula-Santos told me, "One of my hopes is to not just take photographs of food, but also be able to animate it. If you see a sizzling steak in front of you, that's just one way of stimulating more senses." '[25]

So, has there, one might ask, been any reaction to all this focusing on the beautiful in food? Well, I suppose there has, though it only started recently and is still small-scale. You may have heard that various supermarket chains are now selling boxes of wonky fruit. This is undoubtedly a good idea, both to help avoid all that food waste and because, in general, the more beautiful a source of food looks, the less aromatic it tends to be. It is a bandwagon that celebrity chefs like Jamie Oliver are jumping on to as well. Of course, the fact that most

---

* If you need some inspiration in this regard, there is a Tumblr site called 'Someone Ate This', dedicated to awful-looking food. Why don't you check it out, and see if you are still feeling hungry after viewing some of the truly ghastly meals shown there.

consumers reject bruised-looking fruits and unusually shaped vege-tables highlights, once again, the importance of eye-appeal.

Speaking for myself, as the son of a one-time greengrocer I know only too well that the best banana cakes are always made with the ugliest, blackest fruit. The family joke is that us kids never realized that bananas came in any colour other than black, as my father always brought home the leftover produce that no one else wanted to buy, the irony here being that the blackened fruit tend to be the most flavourful.

Seeing sense

Our brains have evolved to find sources of nutrition in food-scarce environments. Unfortunately, however, nowadays, we are surrounded by more images of energy-dense, high-fat foods than ever before. While there is undoubtedly an increasing desire to view images of food, not to mention take pictures of it, and more is known about what aspects of these images attract us than ever before, we should, I think, be concerned about just what consequences such exposure is having on us all. I, for one, am growing increasingly concerned that all this 'digital grazing' of images of unhealthy energy-dense foods may be encouraging us to eat more than we realize and nudging us all towards unhealthier food behaviours in the long term.

Describing desirable images of food as gastroporn, or food porn, is undoubtedly pejorative. However, I am convinced that the link with actual pornography is more appropriate than we think. And given the oft-discussed concerns expressed about the latter, perhaps we really should be thinking seriously about moving those food maga-zines bursting with images of highly calorific and unhealthy food up on to the newsagents' top shelf? Or preventing cookery shows from being aired on TV before the watershed? While such suggestions are, of course, a little tongue-in-cheek, there is a very serious issue here. What is more, the explosion of mobile technologies means that we are all being exposed to more images of food than ever before, presented with foods that have been designed to look good, or photograph well, more than for their taste or balanced nutritional content.

Let me end with a quote from Max Ehrlich's 1972 book *The Edict*, set in a future where the strictly calorie-rationed populace can go to the movies (the Vistarama Theater) to see a 'Foodie': 'To those watching, what they saw was almost unbearable, both in its pain and ecstasy. Mouths dropped half open, saliva drooling at their corners. People licked their lips, staring at the screen lasciviously, their eyes glazed, as though undergoing some kind of deep sexual experience. The man in the film had finished his carving and now he held a thick slice of beef pinioned on his fork and raised it to his mouth. As his mouth engulfed it, the mouths of the entire audience opened and closed in symbiotic unison with the man on the screen . . . The Foodie had been designed to titillate, and it did. What the audience saw now was not simple greed. It was pornographic. Close-ups of mouths were shown, teeth grinding, juice dribbling down chins.'[26]

But I don't want to leave this chapter on a pessimistic note. In the coming years, gastrophysicists will surely continue to examine the crucial part the foods we are exposed to visually play in our food perception and eating behaviours. There seems little chance that the impact of sight will decline any time soon, especially given how much time the majority of us spend gazing lovingly at our mobile devices and computer screens. So my hope is that by understanding more about the importance of sight to our perception of, and behaviour around, food and drink we will all be in a better position to optimize our food experiences in the future.

4.   Sound

Ask yourself which sense is the most important when it comes to determining your experience of food and drink. Most people will mention taste first. Smell will rank pretty highly too, of course. Some might talk about what a food looks like, and maybe even the mouth-feel and oral texture. But virtually no one, be they sensory scientist, chef or regular consumer, talks about sound. However, as you will see in this chapter, what we hear when we eat and drink – even the noises of food preparation, the rattling sounds of product packaging or loud background music – plays a much more important role than any of us realize. Sound, in other words, is the forgotten flavour sense.

The sounds of preparation

What would go through your mind if you were sitting in a fancy restaurant and you suddenly heard the unmistakeable 'ding' of the microwave? It would be pretty disconcerting! I would argue that the sounds of food and beverage preparation are important precisely because they help set our expectations. No wonder, then, that so many people deliberately try to disable the microwave's distinctive sound because of the negative impression that it conveys to all who hear it (especially in a restaurant setting). Go online and you'll be amazed how many blogs and discussion groups there are complaining about this sound and requesting help to turn it off. Large electronics manufacturers such as GE have, in recent years, been working on redesigning it. (Perhaps, though, attitudes are changing, at least in the home environment; according to the results of a recent survey, a third of those questioned said that they wouldn't mind if served a micro-waved meal at a dinner party.[1]) Of course, the sounds of food cooking can also capture our attention. Just think about all of the desirable food preparation sounds that may have you salivating before you can say 'dinner time'. In fact, one of the classic observations in the field of psychology made by the Russian scientist Ivan Pavlov, way back in the 1920s, was that the dogs he was studying started salivating in response to the sound of the bell used to alert them to the arrival of food. The dogs had soon come to associate the sound of the bell with the delivery of food.

Just take the grinding, gurgling, spluttering and sizzling noises emanating from a coffee machine. These sounds are diagnostic, i.e., they are rich in clues about the probable tasting experience that is to follow. Even the screeching and squealing of the hot-air bubbles that make the milk froth provide information, at least for those who know how to listen. It's the change in pitch that tells the barista when the milk in the jug has reached the right temperature. And if you think that's impressive, what about the guy who says that he can distinguish a hundred different brands of beer based simply on the sound of the bubbles when a drink is poured into a glass!

Klemens Knöferle, a former post-doc in my Oxford lab, conducted a study in which he systematically influenced what people said about a cup of Nespresso coffee simply by filtering the sounds made by the machine as it turned those colourful pods into cups of coffee. As he enhanced the harsher, higher-pitched noises, people said that the coffee didn't taste so good. When he cut them, suddenly taste ratings went up. So it's no wonder that so many manufacturers are now trying to engineer the 'right' noises into their machines. They are, in other words, slowly catching up with all the car companies out there who, for decades, have been modifying the design of everything from the sound the car door makes when it closes through to the noise made by the engine as heard by the driver inside the vehicle. Do you remember the iconic Volkswagen 'Sounds just like a Golf' adverts?

Some innovative chefs have started to work creatively with the sounds of preparation. This is what you would have experienced had you been lucky enough to score a table at Mugaritz in San Sebastián in 2015: at one point during the meal, mortars and pestles were brought out from the kitchen, and diners were encouraged to grind their own spices prior to a hot broth being poured into each and every mortar. Just imagine a roomful of diners sitting in a prestigious two-Michelin-starred restaurant, all grinding their spices in synchrony, creating a resonant sound that fills the room. All of the guests sitting at their separate tables united, at least for a moment, by the playful sounds of food preparation.

One Swedish composer, Per Samuelsson, has made something of a career composing with these sorts of food preparation sounds. He is often to be found recording the peeling, chopping, slicing, dicing, grinding, shaking and stirring noises in a busy kitchen. The young Swede then turns those kitchen sounds into a musical composition that is played back to diners while they eat the fruits of the chefs' labour.[2] These compositions are intended to highlight the often unacknowledged effort involved in creating the food that we consume, while at the same time delivering an immersive multisensory environment, one that is designed to enhance the meal experience. Meanwhile, Massimo Bottura, voted the world's top chef in 2016, was recently recorded in an anechoic chamber. The aim? To capture all of the sounds of him making lasagne, the favourite comfort food of his childhood.[3]

It started with a crisp

Back in 2008, Max Zampini and I were awarded the Ig Nobel Prize for Nutrition for our groundbreaking work on the 'sonic chip' (see Figure 4.1). Yes, I know what you are thinking: completely preposterous. Ten of these prizes are awarded every year to a select bunch of international scientists for research that, on the face of it, sounds crazy, ridiculous or preferably both. But the point is that this work, which hopefully makes you laugh first, is actually serious. And getting the prize garners a huge amount of publicity for the winning researchers. Believe it or not, a decade after receiving the award, I am still fielding enquiries on a monthly basis. (This despite my family's incredulity – 'Not the crisps again,' they moan whenever they see the story resurface in the press.) Film crews from around the world

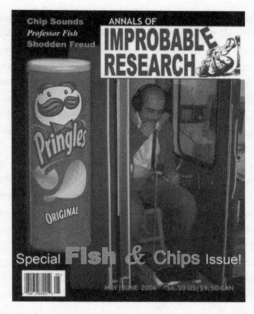

Figure 4.1. My former student Max Zampini (now the esteemed Prof. Zampini) demonstrating the 'sonic chip' experiment on the front cover of *The Annals of Improbable Research*.

periodically descend on the lab, wanting to recreate the magic moment when, simply by changing the sound of the crunch, we were able to change people's perception of the crunchiness and freshness of the crisps (or potato chips, if you are reading this in North America). Truly, my life has not been the same since.

When we first published our results in what, even by academic standards, was a fairly obscure outlet, we thought that they were important, certainly, but nothing out of the ordinary. While Unilever, who funded the study, were interested in our findings, it was hard to see who else might get excited by them. And, if you are wondering why exactly Unilever would fund a project using one of their competitor's products (at the time, Pringles were owned by P&G),

well, the answer is that Pringles are ideal for gastrophysics research. Why so? Because they are *all* exactly the same size and shape. So you can be sure that any change in people's responses must be attributable to the sound manipulation that you have introduced, not to any individual differences in the crisps themselves. And Pringles have another practical advantage: they are large enough that people don't normally put them in their mouths in one go. (You don't, do you?) Hence the relative contribution of air-conducted to bone-conducted sounds in the overall multisensory tasting experience is enhanced.

Basically, we discovered that, solely by boosting the high-frequency sounds that people hear when they bite into a Pringle, we could make them seem around 15% crunchier and fresher than when we cut those sounds. Of course, you might reasonably expect that chefs would be less influenced by such superficial sound modifications of the food they eat. But that turns out not to be the case, at least not if the trainee chefs at the Leith cookery school in London were anything to go by. When we tested them for a BBC TV show a few years ago, they were so busy concentrating on the texture of the crisps that they were just as easily fooled as the Oxford undergraduates who provided the subject matter for our original study.

You can play exactly the same sonic tricks with apples, celery, carrots or, in fact, with any other noisy food, be it dry, like crisps and crackers, or moist, like fruit and vegetables. In one recent study, this time conducted in northern Italy, ratings of the crispness and hardness of three varieties of apple were systematically modified by changing people's biting sounds.[4] This crossmodal illusion is important for a couple of reasons. For one thing, it provides one of the most robust demonstrations that what we hear really does influence what we taste. And it turns out that this particular crossmodal effect works just as well even when you know exactly what is going on. It continues to work no matter how many sonically enhanced chips you've bitten into too. I should know, having crunched more than most – all in the name of science. In other words, the sonic chip illusion is an automatic multisensory effect, as my colleagues in the cognitive neurosciences would say.

Can the sonic chip change product innovation?

But perhaps the more important consequence of our 'ground-breaking' research is that this kind of neuroscience-inspired testing protocol has now been adopted by many of the world's largest food companies. The virtual prototyping approach developed here in Oxford, where we assess how consumers respond to augmented reality products (rather than real product prototypes) allows these companies to figure out how people would respond if their products were to have added crunch or crackle, say. Importantly, the food and beverage companies can do this without having to go into the development kitchens and make a whole host of new products that actually do sound noticeably different from each other. This, the traditional approach, tends to make for a much slower and more effortful development process, especially when the feedback from the tasting panel, as is often the case, is that they didn't like *any* of the new variants that the researchers worked so hard to create. On hearing such news, they have to go back to the kitchens, shoulders slumped and heads drooping, and prepare a whole new set of samples. This is product innovation, yes. But it can be painfully slow!

By contrast, having the testing panel evaluate virtual product sounds first allows one to assess a whole range of alternatives, and find out what, if anything, really makes a difference. So the process is inverted: first, you try to figure out what sounds people like their products to make, and only then do you go into the kitchens to determine whether the chefs and culinary scientists can actually create foods with the requisite sonic properties. Sometimes, those working in the kitchens will shake their heads and laugh, replying that what they are being asked to do is physically impossible. Other times, though, they will know just what is required. But whatever they say, at least everyone knows in which direction they should be heading, sonically speaking. And as a result, product innovation occurs much more rapidly.

The sound of food

Many of the food properties that we all find highly desirable – think crispy, crackly, crunchy, carbonated, creamy* and, of course, squeaky (like halloumi cheese) – depend, at least in part, on what we hear. Most of us are convinced subjectively that we 'feel' the crunch of the crisp. However, this is simply not the case. Introspection, after all, often leads us astray and, based on the results of the gastrophysics research, I can assure you nowhere is this more true than in the world of flavour. (Take, for instance, the experience of carbonation. Most people, if you ask them, will swear blind that they enjoy the 'feel' of the bubbles bursting or exploding in their mouths. It turns out, though, that the sensation is actually mediated largely by the sour receptors on the tongue; i.e., by the sense of *taste*, not by the sense of *touch* at all.)

Given that we don't have touch receptors on our teeth, any feeling we get as we bite into or chew (masticate) a food is largely mediated by what is felt by the sensors located in the jaw and the rest of the mouth. The latter, removed as they are from the action, do not provide any especially precise information about the texture of a food. By contrast, the sounds that we hear when a food fractures or is crushed between our teeth generally provide a much more accurate sense of what is going on in our mouths. So it makes sense that we have come to rely on this rich array of auditory cues whenever we evaluate the textural properties of food.

Some of these sounds are conducted via the jaw-bones to the inner ear, while others are transmitted through the air. Our brains integrate all of these sounds with what we feel and in the case of the sonic chip this happens both immediately and automatically. And so if you change the sound, the perceived alteration in food texture that follows is experienced as originating from the mouth itself, not as a funny sound coming from your ears. This means that most of us are

---

* 'Creamy?' I hear you ask. Yes, even creamy foods subtly change the sound your tongue makes when you run it around your mouth.

oblivious to just how important the sound of the crunch is to our overall enjoyment of food! And this isn't just about the crunch: the same goes for crispy, crackly, creamy and carbonated, though the relative importance of sonic cues to our perception of texture and mouthfeel probably varies for each particular attribute. My suspicion is that what we hear is probably more relevant, and hence influential, in the case of crackly, crunchy and crispy foods than in the case of carbonation and creaminess perception. Nevertheless, research points to the conclusion that what we hear plays at least some role in delivering *all* of these desirable mouth sensations, and more.

Crispy and crunchy

One of the major problems associated with working in this area stems from the fact that despite all of the research that has been conducted over the years, it is still not altogether clear just how distinctive 'crisp' and 'crunchy' are as concepts to many food scientists, not to mention to the consumers whom they spend their life studying. Certainly, judgements of the crispness, crunchiness and hardness of food turn out to be highly correlated, thus suggesting that they are indeed very similar concepts to most of us. But matters start to get more complicated when it comes to languages other than English. Some use different terms; others simply don't have any relevant descriptors with which to capture these textural distinctions, if such there be. The French, for example, describe the texture of lettuce as '*craquante*' ('crackly') or '*croquante*' ('crunchy') but not as '*croustillant*', which would be the direct translation of 'crispy'. Meanwhile, the Italians use just a single word, '*croccante*', to describe both crisp and crunchy sensations.

Things become really confusing when it comes to Spanish. For Spanish-speakers don't really have words for 'crispy' and 'crunchy' or, if they do, they certainly don't use them. Colombians, for instance (and, I imagine, the Spanish-speakers from many other South American countries), describe lettuce in terms of its freshness ('*frescura*') rather than as crispy. And when a Spanish-speaking Colombian wants to describe the texture of a dry food product, they will either borrow the English word 'crispy' or else use '*crocante*' (the equivalent of the French word '*croquante*'). This confusion apparently extends to mainland Spain, for when questioned 38% of consumers there did not even know that the term for 'crunchy' was '*crocante*'. What is more, 17% of the consumers in one study thought that 'crispy' and 'crunchy' meant the same thing. This is kind of bizarre when you think how important noisy foods are to our experience and enjoyment of eating. Given that we still can't seem to agree on the definitions and differences between the various textural attributes of food, it is perhaps little wonder that research on the sound of food hasn't progressed quite as rapidly as one might have hoped. This is unfortunate for, as top chef Mario Batali has noted: 'The single word "crispy" sells more food than a barrage of adjectives describing the ingredients or cooking techniques.'[5]

'Why is a soggy potato chip unappetizing?' This was the title of a commentary in a top science journal called (you guessed it) *Science* a few years ago.[6] The nutrient content doesn't change as a crisp becomes stale but, for whatever reason, none of us seems to like the soggy variant. And yet, no one was, I suspect, born liking noisy food. It is on this point that I have to disagree with Mario Batali when he says: 'There is something innately appealing about crispy food.'[7] No, there isn't. Indeed, most of what we think of as innate is, in fact, learnt. In other words, we all *learn* to like specific food sensory cues, in large part because of what they signal to our brains about what we are consuming (and what physiological rewards are to come). Crisp and crunchy – well, they signal fresh, new and maybe seasonal too.

Perhaps the more fundamental question that we should concern ourselves with is why exactly crispy, crunchy and crackly have come to constitute such universally desirable food attributes? These sounds don't directly signal the nutritional properties of a food – or do they? Let's examine crunchiness, which undeniably provides a pretty good (i.e., reliable) cue to freshness in many fruits and vegetables. This information would have been important for our ancestors, since fresher foods preserve more of their nutrient content and hence are better to eat.

Cooking induces the Maillard reaction, the non-enzymic browning that results from strong heat being applied to nitrogen- and carbohydrate-containing compounds. In his book *The Omnivorous Mind*, John S. Allen points to the fact that cooking by fire simultaneously makes foods both more nutritious (or, more accurately, easier to digest) and crispier. Think, for instance, of the delicious crustiness of freshly baked bread. Hence this may be why, in evolutionary terms, the sounds of crispy, crackly, crusty and crunchy are so important. Part of the answer to the question of why soggy foods are so unappealing may also relate to the latest research showing that as the crunchiness of a food increases, so too does its perceived flavour. No surprise then that we all want more crunch. In fact, consumers the world over demand it!

Our strong attraction to fat may also be relevant here. The latter, after all, is a highly nutritive substance, perhaps explaining why we have taste receptors in the oral cavity sensitive to the presence of fatty acids. However, it can still be a challenge for our brains to detect fat directly in food and drink. Why so? Well, often its presence is masked by other tastants like sweet and salt. While cream, oil, butter and cheese are all associated with a distinctively pleasing and desirable mouthfeel, my suspicion, at least as far as dry snack foods are concerned, is that our brains may have learnt, as a result of prior experience (i.e., exposure), that these sonic cues are suggestive of the presence of fat. That is, the louder the crunch, crackle, etc., the higher the fat content of the food we are biting into is likely to be. So we all come to enjoy those foods that make more noise because they probably have more of the rewarding stuff in them than other, quieter foods. Now you know why you find it so hard to resist the sound of crunch!

How do you feel about eating insects?

I'm sure I know the answer to that question without you having to say a word. But, knowing what we now know about sound, we should all find the idea rather more desirable than, in fact, we do. Many insects are, after all, pretty crunchy, at least those with a hard

Figure 4.2. A crunchy and protein-rich snack. Deep-fried crickets should tick all the right boxes as a highly desirable and moreish treat.

carapace or exoskeleton (see Figure 4.2). What's more, they provide an excellent source of protein and fat. It would be good for the planet, too, if we all ate more of these little critters (and less red meat). And yet, no matter what you or I say, most Western consumers are yet to be convinced. Marketing insects so that people – by which I mean us in the West – find them desirable is, I feel sure, going to be one of the ultimate challenges for gastrophysics in the coming years. How exactly do we take everything we know about the mind of the consumer and make this currently most undesirable food source truly delicious? Or, at the very least, make sure that insect matter constitutes a larger proportion of our diets. Playing on the sound of crunch might offer one way in to the popularization of entomophagy (that's the fancy term for eating insects).

So what gastrophysics insights might help here? Well, one option is simply to surreptitiously increase the amount of insect matter that people already eat. (If you are a peanut butter fan, you might want to skip the rest of this paragraph!) I bet you didn't know that there can be as many as 100 insect parts, for example, in every jar of peanut butter before the producer has to declare this on the label). The same must presumably also be true for jam (due to the difficulty of keeping tiny creatures out), and who knows what else is in your ground coffee. So why not slowly increase that number (while at the same time reducing other ingredients that are in short supply, or are unhealthy)? I bet that in the future insect-based matter will become a more substantial part of our diets without the consumer ever noticing. This is like the health-by-stealth strategy that worked so well for the cereal companies when they decreased the amount of salt in their breakfast cereals. They reduced the salt content by as much as 25% by doing it gradually. Consequently, each successive reduction was essentially imperceptible to the consumer; and over the long term, levels of unhealthy ingredients have dropped substantially.

Alternatively, we might work on distinguishing between more and less disgusting critters. For, if you think about it, we already eat lots of bee-based products, everything from honey (sometimes mistakenly called 'bee vomit') through to royal jelly and propolis. So eating bee brood (i.e., baby bee) ice cream shouldn't be too much of a leap, should it? We don't seem to find ladybirds so disgusting either – certainly if one lands in my beer, I will happily flick it out and keep on drinking.

It may be, though, that the sensory strategy that works best in the long term is one that builds on the sound of crunch – something, after all, that we know most consumers really like. All the gastrophysicists have to figure out now is which insects, and which method of preparation, would make the loudest crunch of all? Then away we all go, to a crispier, crunchier and more sustainable future.

Why do crisps come in such noisy packets?

As well as the sounds of preparation and the noises associated with our consumption of food and drink, the sounds of product packaging also have a pronounced impact on our tasting experiences. Do you think that it is an accident that crisps come in such noisy packets? Of course it isn't! From the very beginning, marketers intuited that it would make sense to have the sound of the packaging be congruent with the sensory properties of the contents. This is as true today as it was back in the 1920s when crisps were first packaged for fresh, portion-controlled delivery direct to the consumer. Even Pringles, whose packets typically make less noise than most other snacks, have done something to enhance the sound of their foil seal. You don't have to take my word for it – try running your fingers over it next time you come across a tube and just listen to the difference.

But just how much influence does the sound of the packaging *really* have on our judgements of the product within? Well, a few years ago, we tackled this very question. Together with Oxford undergraduate Amanda Wong, we conducted a study showing that the louder the rattling packaging sound that people heard as they ate, the crunchier-seeming were the crisps they had been asked to rate. While the effects were nothing like as dramatic as those we saw when we modified the sound of the crunch itself, they were still significant. In other words, in terms of perception, our brains appear to have a remarkably hard time distinguishing the product from the packaging.

Frito-Lay may have taken these findings a little too seriously when they came out with their all-new biodegradable packaging for SunChips. This, let me tell you, is probably the loudest packaging ever created (see Figure 4.3). When my colleague Barb Stuckey sent me a couple of packs from California, we got the sound meter out to determine in the lab how much noise they gave off when gently agitated in the hands. The answer: in excess of 100 dB! To put that number into perspective, this is the background noise level that you will find in the very loudest of restaurants. Moreover, it is the sort of noise level that, if one is exposed to it for long periods of time, can lead to permanent hearing damage!

The packaging was so loud that many consumers wrote in to complain. It was so loud that the company was forced to offer ear plugs to try and quell the growing consumer backlash. The idea, I suppose, was that you would buy a packet of SunChips, get them home, then insert the ear plugs and enjoy your snack in peace. You'd have to feel sorry for anyone who was sitting nearby. Even worse if they suffered from misophonia (sufferers find the sound of other people masticating unbearable). Eventually, of course, this damage-limitation exercise failed miserably, and the company withdrew their

Figure 4.3. Sonic warfare? Quite possibly the noisiest packaging on earth, creating in excess of 100 dB of rattling noise when the packaging is gently agitated in the hands. Who thought that was a good idea?

sonically supercharged packaging from the shelves altogether, never, I suspect, to be seen nor, more importantly, heard again.

You can almost hear the marketing executives rubbing their hands together with glee, though, at the ingeniousness of their original idea. We had shown at the Crossmodal Research Laboratory that noisy packaging helps bring out the crunch of the crisp, so surely selling your noisy product in even noisier packaging ought to make your product seem crunchier still. (Assuming, that is, that people eat direct from the pack, rather than pouring the contents out into a bowl or plate. This, though, is a reasonable assumption, given claims that a third of all food is eaten direct from the packaging; if anything, I would imagine that this figure is likely to be much higher in the case of crisps.) The noisy packaging also had the advantage of being extremely effective at capturing the attention of consumers. For as soon as anyone picked a pack off the shelf in the supermarket, you could guarantee that everyone else in the aisle would be looking around to see what the hell was going on. While I doubt there will be anything as noisy ever again, you may be sure that many other companies are thinking about or even actually subtly modifying the sounds that their packaging makes. But the real moral of the SunChips story is 'everything in moderation'. Just because more sound is good, that doesn't mean that even more sound will necessarily be better!

'Snap, crackle and pop'

Product sounds can help set our expectations regarding the product category, or even the specific brand. Some years ago, Kellogg's even tried to patent their crunch. They wanted to trademark the distinctive sound that their breakfast cereals made when the milk was added. They hired a Danish music lab to create 'a highly distinctive crunch uniquely designed for Kellogg's, with only one very important difference from traditional music in commercials. The particular sound and feel of the crunch was identifiably Kellogg's, and anyone who happened to help himself to some cornflakes from a glass bowl at a breakfast buffet would be able to recognize those anonymous cornflakes as Kellogg's.'[8]

Packaging sounds can help set our expectations too. According to Snapple (the beverage company owned by Dr Pepper Snapple Group, Inc.), the distinctive (or signature) sound that the consumer hears on unscrewing the cap from an unopened bottle provides a cue to freshness. 'The company calls it the "Snapple Pop" and says it builds anticipation and offers a sense of security, because the consumer knows the drink hasn't been opened before or tampered with. Snapple was so confident about the pop's safety message that in 2009 it eliminated the plastic wrapping used to encircle the lid. It saved on packaging costs and eliminated an estimated 180 million linear feet of plastic waste, the company says. "We were a lot more comfortable making that decision because we knew there was this iconic pop," says Andrew Springate, senior vice-president of marketing.'[9]

Given how much money the food and beverage companies spend on developing and protecting their visual identity, it is surprising how few of them seem to give much thought to their sonic identity. Interestingly, though, as far as I am aware, Snapple hasn't protected the 'Snapple Pop'. This may be because it has proved hard to trade-mark specific product sounds (as Harley-Davidson found to their cost when they tried to protect the rich, low-pitched 'potato-potato-potato' sound of their motorcycle exhausts).

Many advertisers have picked up on the potential of sound – not much escapes the ears of marketers. Often they attempt to draw attention to the noises that products make when opened, poured and/ or consumed on screen. The JWT ad agency, for example, worked on a campaign in Brazil to emphasize the sound that Coca-Cola makes when poured into an ice-filled glass. Or think of the loud crack you hear on the Magnum ice cream ads, or the iconic sound of someone running their fingers along the foil seal of the old-fashioned Kit-Kat.* What other distinctive food packaging sounds can you think of?

What does your food sound like at home?

All right, I hear you say, I can see why big companies or chefs might be interested in sound or in sonically mediated food textures, but how does this affect us *mere mortals*? Well, the latest findings from gastrophysics highlighting the importance of sound also provide insights that you can take advantage of at home. For instance, the next time you throw a dinner party, be sure to ask yourself where the sonic interest lies in the dishes you serve. If it isn't crunchy, crackly, crispy or creamy, are you stimulating your guests' senses as effect-ively as you might? The solution can be quite simple: just sprinkling

---

* This latter is just one of those rituals, as targeted by questions like 'How do you eat your Oreo cookie?', that has been chucked out by the bean-counters who, wanting to save money on packaging costs, changed Kit-Kat wrappers from foil and paper to plas-tic. Shame!

some toasted seeds over your salad, or adding some crispy croutons to
your soups at the last minute. This presumably explains the ubiquitous presence of the gherkin and Batavia lettuce (also known as French
crisp lettuce) in your burger bun too – they add a sonic element that
makes you enjoy the experience of eating the burger that much more.

Those of you who are a little more adventurous might want to try
sprinkling some popping candy into your chocolate mousse, or even
into the potato topping of your shepherd's pie. These are both
approaches that top chefs have incorporated into their dishes over
the years. And if you want to make the sonic surprise all the more
memorable, 'hide' it. Your guests will be taken aback when, several
mouthfuls into that mostly silent chocolate mousse, say, they suddenly experience an explosion of sound in their mouths. Something
that they won't soon forget, I can assure you! This is a good thing, as
we will see in the 'Meal Remembered' chapter.

Have you ever wondered why the pairing of Melba toast with pâté
works so well? Is this not a classic example of taking a great-tasting
but silent food (that's the pâté), and adding to it a burst of noise (when
biting into the crisp toast)? Sure, there is texture contrast here (and
that is important too). But fundamentally, is it not really about injecting some sonic interest into the dish? More often than not, in fact, our
perception of flavour in a meal is enhanced when sound is introduced.
Indeed, as we saw earlier, some of the most interesting recent research
has shown that as the crunchiness of a food increases, so too does its
flavour. In *The Omnivorous Mind*, J. S. Allen also suggests that loud
foods might be more resistant to habituation than silent ones. This, he
believes, might be part of their universal appeal. So there you have
it. Whatever you do, make sure you add some sonic excitement to
mealtimes. Better just first check that there aren't any misophones at
the table . . .

If sound is as important to the perception of crunchiness and crispness as we think, then maybe there is a solution here if you find that
you only have stale crisps in the cupboard next time you throw a
party. According to research, your guests probably won't realize just
as long as you turn the background music up loud enough to mask the
missing sound of the crunch. For as soon as you introduce some loud

background noise, your guests' brains will fill in the missing sound of the crunch that they can no longer hear properly. Be warned, though: all that loud noise is also likely to impair your guests' ability to determine how alcoholic your punch is. And if any of them should be impertinent enough to ask you why the music is so loud, just tell them that all the best chefs do it nowadays (about which more below).

'Pardon?' Taking the din out of dinnertime

When was the last time you went out to eat, only to find it difficult to hear the conversation of your companions? My guess is that it probably wasn't long ago. The problem of overly loud restaurants (not to mention bars) has become far more widespread in recent years. Public spaces can be so noisy that many of us can no longer hear ourselves think, never mind get our orders across. Noise is currently the second most common complaint amongst restaurant-goers, behind poor service. In fact, over the last decade or two, many restaurants have become so loud that some critics now report on the noise levels alongside the quality of the food in the venues they judge.

The blame for this growing cacophony has been laid, in part, at the doors of the New York chefs, famous for listening to very loud music while preparing for service. No one knows quite when it started, but at some point one of them had the 'bright' idea that maybe the diners would like it as well. Bad move! For as one journalist astutely notes: 'No matter how elegant the food at a restaurant, the music that plays as it's prepared is likely to be less refined. No one is listening to Vivaldi as he buffs baby vegetables and dismembers ducks.' Charles Michel, the former chef-in-residence at the lab, told me that when he worked at the Louis XV restaurant in Monaco's Hôtel de Paris, the chef de cuisine, Frank Cerutti, would put on some heavy metal during the *mise en place* in order to make the kitchen staff go faster![10]

Part of the responsibility must lie with those designers who have insisted on the restaurants they advise ditching their soft furnishings, leaving spaces filled with hard reflective surfaces. The new Nordic look – you know, all that bare wood, and with carpets, upholstered chairs and tablecloths removed – has a lot to answer for, sonically speaking! There is nothing left to absorb all the noise. That said, though, the chefs might not be entirely blameless. After all, part of the reason for removing the tablecloths in Grant Achatz's Michelin-starred restaurant Alinea, in Chicago, was to give the dishes a little more sonic interest when first placed on the table.

The backlash against restaurants and bars that are overly noisy has been growing steadily. According to a recent press report: '[A] group of Michelin-starred chefs have started a campaign to reduce noise levels in Spanish restaurants amid concerns that it is spoiling the gastronomic experience for some guests.' Others have gone further. Ramón Freixa, chef at the two-Michelin-starred Hotel Único in Madrid had this to say recently: 'Gastronomy is a sensual experience and noise prejudices this pleasure. A good conversation with the people you are dining with should be the only sound in a restaurant.'[11] So how is the din to be taken out of dinnertime? While a successful restaurant full of customers can do without music, it is obviously going to be a much harder trick to pull off in a lightly pop-ulated venue. Furthermore, it is also worth remembering here that background music serves to insulate one table's conversation from the prying ears of those seated nearby. A better aim, then, might be to set the volume level so that the music is 'very present but it's never overpowering.'[12]

A number of chefs and culinary artists have spotted a gap in the market and have been arranging silent dining events. If you eat in silence, then you can really concentrate on the food (and no, texting isn't allowed either, before you ask). This should enhance the sensory pleasure of the experience. Much the same logic underpins all those dine-in-the-dark restaurants out there. Such mindful dining might even lead to reduced consumption. That said, these events have not been a commercial success, I suspect because while this strategy can help to build anticipation by emphasizing the sounds of preparation coming from the kitchen (at least if managed appropriately), insisting on silence also precludes the main activity when eating and drinking, namely the social aspect of communicating with those who you are with.

Across a number of studies we, and a growing number of other researchers, have been able to demonstrate that what we listen to (and how much we like what we hear) influences the taste, texture and aroma of a diverse range of foods. People's hedonic ratings (i.e., how much they like the food or drink) are also affected. Often – but, interestingly, not always – the more we like the music, the more we enjoy the taste of the food or drink consumed while listening to that music. For instance, one recent study revealed that liked music brought out sweetness in gelati, whereas disliked music tended to bring out bitterness in a group of trained panellists. While some have wanted to dismiss such results as little more than a party trick, I would argue that this research may actually have some profound implications for the way in which we think about the senses and food. Furthermore, such results will, I believe, affect the future design of multisensory tasting experiences, no matter where you may happen to find yourself. In fact, we will come across a number of examples in the chapters that follow. So my prediction is that you are going to hear a lot more about sonic seasoning in the years to come.

Sonically enhanced food and drink

When it comes to consumption, many of the food attributes that we find most pleasurable – think here of the crispy, crunchy, crackly, creamy and carbonated – are all influenced, to a greater or lesser extent, by the sounds we hear during consumption. While this is undoubtedly important for younger consumers, in the future, I suspect, boosting the sonic interest of one's dishes is going to become even more important for the rapidly growing ageing population. After all, the number of us living past seventy is steadily increasing, and this is the age at which one starts to see some really dramatic decline in the ability to taste and smell (much though my octogenarian parents deny it!). Now, for any of you out there feeling relieved that this doesn't apply to you, I'm afraid that I've got some bad news.

According to the sensory scientists, if you are out of your teens, the chances are that the decline has already started. Not feeling quite so

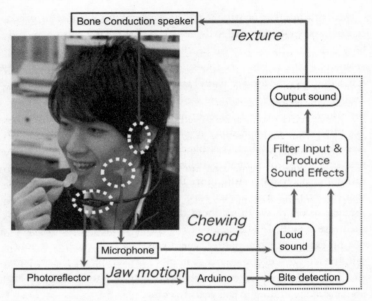

Figure 4.4. The playful 'Mouth Jockey' detects the user's jaw movements and then plays back a specific pre-recorded sound.

smug? I didn't think so. Research clearly shows that the majority of older adults are pretty much anosmic: i.e., they can no longer smell anything much at all. Unfortunately, there is currently absolutely nothing that can be done to bring back the sense of taste or smell once they have begun their inevitable decline (unlike the boost we can give failing vision or hearing with spectacles or hearing aids). But one thing that can be done is to make sure that dishes served to the elderly have lots of crunch and crackle – more sonic interest, in other words. This should help to stimulate the mind and palate of whoever is eating.

So what have younger people, whose taste buds and, more importantly, noses are still in good working order, got to look forward to? Researchers in Japan have developed a playful headset called the 'Mouth Jockey' (see Figure 4.4), which detects the user's jaw movements and plays back pre-recorded sounds while the user eats. Just imagine hearing the sound of screaming as you bite into a gummy bear, say. Others have been working on augmented straws that can recreate the sounds and feeling associated with sucking liquidized food through a straw: place the straw over a mat showing a picture of the desired food and then suck. It is amazing how realistic and how much fun the experience can be.

The EverCrisp sonic app allowed you to 'freshen up' your stale crisps by using your mobile device to add a little more crunch. While it is tempting to imagine that technology will come to play an increasingly important role in augmenting our experience of food in the years to come (see the 'Digital Dining' chapter), I believe that there will also be a crucial role for design. I want to leave you, therefore, with the Krug Shell (see Figure 4.5), as an example of what is possible here. This Bernardaud Limoges porcelain listening device, which was released as a limited edition offering back in 2014, sits snugly on top of a custom Riedel 'Joseph' champagne glass. If you can somehow get your hands on one, I'd encourage you to try it. You will hear how the sounds of all those bubbles popping in your glass have been pleasantly amplified. Then sit back and contemplate whether you are really happy to let sound remain 'the forgotten flavour sense'.

Figure 4.5. The Krug shell, designed by French artist Ionna Vautrin to amplify the sound of the bubbles popping in a glass of champagne.

5. Touch

At a number of the world's top Michelin-starred restaurants these days, the first three, four or even five courses are all eaten with the fingers. This is happening at Noma in Copenhagen, at Mugaritz in San Sebastián, Spain, and even at The Fat Duck in Bray. Any self-respecting gastrotourist will know what I am talking about. But just think about it for a moment. This would have been unheard of in two- or three-Michelin-starred restaurants even a few years ago. Sure, we probably always eat with our hands while snacking in the car, and we mostly take the bread with our fingers, even in the fanciest of restaurants, and shellfish too for that matter, but to eat other dishes with our fingers? That is something new. And, going one stage further, Mugaritz declared in 2016 that they were no longer going to use traditional cutlery at all.

In this chapter, I want to show you just how important what we feel is to our experience and enjoyment of food and drink. Not only what we feel in our mouths but also what we feel in the hand. Mark my words, dining and drinking are slowly but surely becoming much more tactile activities. And I am not just talking about the increasing range of textures and mouthfeels that you might be exposed to while eating the latest creation of some famous modernist chef or other; rather, I am talking about our total tactile interaction with food and drink. Touch is, after all, both the largest and earliest developing of our senses, with the skin accounting for about 16–18% of body mass. Ignore it at your peril!

Does texture influence taste/flavour?

The answer is a very definite yes, although this is a hard subject to study empirically, as it is difficult to manipulate the sensory cues independently. So, for example, increasing the viscosity of a liquid will reduce the number of volatile aroma molecules released from the drink's surface. So while you may be certain that the oral-somatosensory properties of a food or drink (basically the mouthfeel) will affect the taste/flavour of food and drink, determining the cause of that interaction can be a little harder to discern.

Not so long ago, though, sensory scientists finally managed to figure out a way of varying texture and aroma independently. Specifically, they delivered the latter by means of a tube placed into the mouth of their unfortunate subjects and were able to demonstrate that the addition of certain fatty aromas modified the 'mouthfeel', and perceived thickness, of liquids. On the other hand, increasing the oral viscosity of the liquid in the mouth (just think about the different mouthfeel associated with cream versus water) also affected the perceived aroma/flavour. This is true even when the aroma is delivered via a tube, and so one can be sure that the aroma release hasn't changed as a result of the change in viscosity.

Why not try eating something like a strawberry or a cookie, or whatever you have at hand? Ask yourself where the taste/flavour originates. The answer, most likely, is that it appears to come from the food that you can feel moving around in your mouth. Am I right? However, as a food breaks down and is transported around the oral cavity via the saliva as a result of chewing (what my colleagues like to call mastication), you are probably tasting with *all* your mouth – and every time you swallow, with your nose too (via the retronasal olfactory route). Your brain does an amazing job of putting all of these sensory cues back together again and associating them in your mind with their presumed source, i.e., the food you feel in your mouth. Such a good job, normally, that we rarely, if ever, stop to think twice about it.

When you watch a film at the cinema, your brain ventriloquizes the voices from the loudspeakers located around the auditorium so that they appear to be coming from the lips that you see moving on the screen. Much the same thing happens whenever we eat and drink. The simplest way to illustrate this phenomenon is to take a teaspoon of a salty or sweet solution into your mouth and then run a tasteless

Q-tip across your tongue. You should feel the taste coming from the tactile stimulus moving across your tongue, despite the fact that the swab itself has no taste. What this shows is how tactile stimulation in the mouth 'captures' where we perceive taste to originate from. In fact, knowing that the brain does this has allowed us (in some of our more bizarre research) to mislocalize taste out of the mouth and on to a butcher's tongue (or a rubber imitation of a human tongue)! Some people are convinced that they can taste what they see being applied (e.g., a drop of lemon juice) to a fake tongue outside of their body when, in fact, what is dropped on to their own tongue is water.

'Oral referral' describes the phenomenon of experiencing food attributes like fruitiness, meatiness and smokiness in the mouth, rather than in the nose, where such aromas are first sensed. For more than a century, it was thought that the tactile stimulation experienced in the mouth while eating and drinking explained the oral referral of odours to the mouth. However, this turns out not to be the case. The fact that the multisensory integration of flavour cues occurs so effortlessly, and so automatically, shouldn't hide how incredibly complex it all is. So, in response to the question: 'Can you feel the taste?', the answer is: 'No.' That said, what we feel, in the mouth or out, most certainly does affect both the taste and flavour of what we eat and drink.

Marinetti's tactile dinner party

F. T. Marinetti, the founding father of the Italian Futurists, had a great interest in touch. He produced a tactile manifesto called *Il tattilismo* in 1921 and organized the very first tactile dinner parties back in the 1930s. Unfortunately, however, there was just one problem: the Futurists couldn't cook. They were dismissed as 'a fart from the kitchen' by the Italian press.[1] Their approach to cuisine was never meant to last (their questionable political views didn't help their cause much either). No wonder, then, that they disappeared without trace in the 1940s. Though, as we will see later in this chapter, contemporary chefs and culinary artists are starting to revive a number of Marinetti's marvellously mad ideas about experience design (forget the food), with some surprising results.

Not being able to cook certainly didn't stop the Futurists from throwing some of the most influential dinner parties. According to a description of one such dinner, the meals were to be eaten to the accompaniment of perfumes which were to be sprayed over the diners, who, holding their fork in their dominant hand, would stroke with the other some suitable substance – velvet, silk or emery paper. So, if you want to give your dinner guests the full Marinetti experience, insist that they turn up wearing pyjamas made of different materials, such as velvet or silk. Then, when the food is placed on the table, encourage them to eat with some kind of textured implement, while feeling their next-door neighbour's jim-jams with their other hand! And, if that seems a tad risqué for the suburbs, I have a couple of more modest solutions that all you budding Futurists out there could try instead.

You could, for instance, take inspiration from London chef Jozef Youssef, of Kitchen Theory. He served a dish entitled 'Marinetti's Vegetable Patch' as part of his sell-out 'Synaesthesia' dining events in 2015. There were various different textures on the plate itself. But there were also a number of black cubes (designed to fit in the hand) scattered across the dining table, each surface, or opposing pair of surfaces, of which was covered with a different material: Velcro, velvet, sandpaper, that sort of thing. The diners were encouraged to taste the various elements of the dish while feeling the textured cubes. They were instructed to look out for any correspondence between what they were feeling in their hands and the textures in their mouths. Now, it has to be said, some people had no idea what was going on. This certainly wasn't one of those dishes/experiments that works on absolutely everyone. But there were some diners (roughly a third, I'd say) who volunteered that their experience of the food had been altered by changing the surface they touched. Weird, almost synaesthetic, some would say. So maybe the Futurists were on to something after all!

Together with Professor Barry Smith of the University of London, we do something very similar when we run tutored multisensory wine-tastings. We get a selection of swatches cut from different materials and hand them out to everyone. We then serve a couple of reds and have people rate how well the different textures go with each of the different wines. This is something so simple, and yet it intrigues many people. And it is something that anyone can try at home. Why not try it yourself next time you open a bottle with some friends? It will, at the very least, get your guests to pay a little more attention to the tasting experience. It may also help to explain why all those adverts for red wine invoke textural metaphors: velvety, silky, and so on.

Is the first taste really with the hands?

In many parts of the world – think Africa, the Middle East and the Indian sub-continent – people mostly eat with their hands. However, in the restaurant setting, especially in Westernized countries, we nearly always eat with the aid of cutlery, be it cold, smooth knife and fork in the West or chopsticks in the East. And whenever we drink, we always pick up the cup, glass, can or bottle first.* In a very real sense, then, the first taste is with the hands. According to sensory scientists and flavour chemists, the feel of the cutlery or drinking vessel shouldn't exert any influence over the taste of the food and drink. Nor should it really affect how much you enjoy the tasting experience. After all, everyone – chefs, food critics and regular consumers alike – thinks that we can simply ignore 'the everything else' and concentrate squarely on the taste of the food on our plate, or the flavour of the drink in our glass. But we cannot! By this stage of the book, I hope you are convinced that 'the everything else' really does matter. And what we feel is no different. In truth, it has more of an influence on our experience of food and drink than many of us would credit, or even, perhaps, be willing to believe.

A growing body of research from the emerging field of gastrophysics now demonstrates how what we feel really can influence our

---

* Except if we drink through a straw. But that is a very bad idea, given that it minimizes the orthonasal olfactory hit! Just see the 'Smell' chapter.

tasting experiences. Some of the world's top chefs, molecular mixologists, culinary artists and even packaging and cutlery designers are starting to pay much closer attention to what we feel when we eat and drink. They are playing around with changing everything from the texture and weight, through to the temperature and firmness of whatever it is that we happen to be holding in the hand while consuming. And take note: they are not stopping at the hands! The most creative experience designers out there are also thinking about how best to stimulate your lips, and even your tongue, more effectively.

Do you really like the feel of cold smooth metal?

We did not evolve to find the cold smoothness of stainless steel or silver cutlery particularly appealing to the touch. Rather, we were always meant to eat with our hands. So why are so many of our interactions with food mediated by metal cutlery? As top interior designer Isla Crawford once put it: 'Surfaces made from natural materials are often preferable, as irregularity is far more sensual than clinically perfect surfaces.'[2] To my way of thinking, it's bizarre: so many of the world's top chefs are doing such amazing things on the plate (assuming there is a plate – not always guaranteed these days), expressing culinary genius and creativity in ways we have never seen, nor could even have imagined back in the 1970s. And yet those self-same chefs have their diners eat with the traditional combination of knife, fork and spoon. It really isn't very imaginative now, is it?

There are, after all, few other objects that you would put in your mouth after they had already been in who knows how many other people's mouths beforehand. How would you feel if I suggested you use someone else's toothbrush? So what, exactly, is so different about cutlery?

In the years to come, I believe, we are going to see some radical innovation in the way in which we move food from plate or bowl to mouth. I hope that open-minded cutlery makers out there will take the scientific insights about the receptor profile of the average human mouth, and all the latest gastrophysics research, and translate this knowledge into aesthetically pleasing cutlery designs that will enhance our tasting experiences. The likelihood is that the results of their labours will first be found within the confines of the modernist restaurant. And from there, signature cutlery will slowly start to appear in the marketplace, perhaps under the brand name of one top chef or another.

Would you enjoy eating with a textured spoon?

So, to get us started on this tactile journey, take the example shown in Figure 5.1. How do you think your experience would be if you ate something with the aid of one of these spectacular-looking utensils? More memorable, probably. More stimulating, absolutely.

Unfortunately, the spoon shown in Figure 5.1 is one of a kind. I very much doubt that you will find this designer's work for sale on Amazon any time soon. A shame, really, given how boring the texture of most cutlery is currently. But at least one mainstream cutlery manufacturer has recently brought out a commercial range of sensorial spoons with added textural interest (see Figure 5.2). These four textured spoons caress your tongue in unusual ways, but we are still

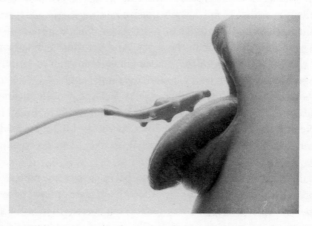

Figure 5.1. 'Tableware as sensorial stimuli cutlery' from the wonderful designer Jinhyun Jeon.

Figure 5.2. A set of four textured spoons from Studio William.

researching, together with chef Jozef Youssef and the top cutlery designer William Welch, into whether any of the spoons do an especially good job of enhancing a specific taste, flavour or texture of food.[3]

For those of you who haven't yet acquired your own set of sensorial spoons, there are nevertheless some very simple ways to stimulate your guest's tongues more effectively without having to invest in a whole new set of cutlery. Next time you invite them round for dinner, why not surprise them? Wet some spoons in lemon juice (best not try this with the silver ones, though, if you don't want to get into trouble). Next, dip them into something crystalline or gritty, like sugar or a little ground coffee, say, and allow them time to dry. Then, just before you are about to serve, place a dollop of something tasty on top and hand them to your guests. This, anyway, is the technique that some of my culinary artist friends, like Caroline Hobkinson, use to tickle their guests' tongues in ways that they might not have been stimulated previously. Even top restaurants have used this approach, for example, Alinea's 'Osetra' dish. At the very least, the unusual texture will surprise the diner and hence make them a little more mindful about what they are eating.

Another way in which to alter your guests' experience of their food is to change the material properties of the cutlery itself. One cheap solution here involves laying the table with some wooden

Figure 5.3. High-end wooden cutlery, as served at Noma in Copenhagen. An unusual texture, yes, but too light for some.

picnic cutlery rather than your normal knives and forks – it'll probably save on the washing-up too! But bear in mind how one diner responded when Noma, currently one of the world's top restaurants, tried something similar. They introduced some high-end wooden utensils into the service at their restaurant in Copenhagen (see Figure 5.3). While I have not yet had the opportunity to try them for myself, one colleague who visited the restaurant in 2015 was certainly disappointed. As she wrote to me after her return: 'It was like eating with a spork from a takeaway.' It would be really interesting to conduct the appropriate study in a restaurant setting, to see whether this response would be true more generally.

Weight, what is it good for?

I cannot emphasize enough just how important weight is to the design of cutlery. You definitely want it to have a good heft in the hand, not to mention the right balance between one end and the other. One of the first things I noticed at Heston Blumenthal's The Fat Duck restaurant was how heavy the cutlery is (really hefty pieces, made from wood and steel, in the French Laguiole style).* The cutlery maker William Welch (the designer of the textured spoons we saw in Figure 5.2) knows intuitively that making sure his cutlery feels good in the hand is really important. As important, he told me, for most people as its looks.

In contrast, it is amazing to me how many young chefs skimp on their cutlery. It is understandable, of course. Imagine some Young Turks opening their first gastropub out in the countryside somewhere. They have poured their life savings into the endeavour, and are likely to be running short of cash when they open. Heavy cutlery might seem like more of a luxury than a necessity, right? But if they cut that corner, they'll end up serving beautifully prepared food to be eaten with what feels like light canteen cutlery. It really detracts from the overall experience, as I am sure you'd intuitively agree. But what does the gastrophysics research say?

Given the differing priorities out there, it was obviously time for the gastrophysicist to step in and conduct an appropriate study. Of course, our first job was to check the scientific journals to see what had already been done. The really surprising thing was that the literature was pretty much silent on the topic (i.e., on the impact of cutlery on the experience of food and drink). How could something so fundamental have been ignored for so long? So, in our own research, we wanted to determine once and for all just how important the weight of cutlery in the hand really is to the experience of food in the mouth (or mind). We had already conducted a series of studies here at the Crossmodal Research Laboratory, demonstrating that if people tasted food with a heavier spoon they generally had better things to say about it than when exactly the same food was eaten with a lighter spoon instead. But evaluating store-bought yoghurt from a heavy plastic spoon in the lab is a long way from the setting of a high-end restaurant. Would the same results also apply *there*?

In many of our highly controlled laboratory experiments, the same people taste putatively different foods with spoons varying markedly in terms of their weight. One of the benefits of using the same participants is that we can be sure that the results we obtain are

* At least it was before the restaurant closed to refurbish their kitchens in 2015.

due to our experimental manipulation and not to individual differences between people. However, on the negative side, it's possible that the format of our studies might have focused the participants' attention unnaturally on weight. Imagine yourself being asked to taste food over and over again, for up to an hour, when the most salient thing that varies is the weight of the cutlery that you are given to use. With nothing else to occupy your mind, there is a very real danger that the cutlery's weight might capture your attention, and hence influence your behaviour, in a way that it simply wouldn't do in a restaurant, say.

Given such concerns (which, it should be said, apply to much laboratory-based research), I was looking out for an appropriate opportunity to test the cutlery idea out in the wild. Of course, I would love to carry out a study with the diners holding that oh-so-heavy cutlery in The Fat Duck. But that is just never going to happen. Why not? Well, which funding body would agree to pick up the tab for all my *subjects* (yes, those well-fed guinea pigs) at the end of the night? Currently the price at the restaurant is nearly £300 a head, and that's before wine or service.

Luckily enough, though, at around this time, I was invited to give a talk at an International Egg Confederation conference. The organizers wondered whether I would mind awfully running a three-course experimental lunch for the delegates, so that they could get an idea of what real gastrophysics research was like. I couldn't believe my luck! They had just offered me the perfect opportunity to test out the theory that weight in the hand really does matter to our enjoyment of food in the mouth. It was the moment that I had been waiting for. But would the experiment actually work, out there in an ecologically valid testing situation?

Imagine the scene: 150 international conference delegates in a swanky hotel restaurant somewhere in the centre of Edinburgh. The diners were randomly dispersed across different tables. There was a scorecard by each place setting and pencils on the tables so that the diners could record their responses. They were asked how much they liked the food, how artistically they thought it had been plated, and how much they would have been willing to pay for the dish in a restaurant like the one they happened to be sitting in. While the conference delegates were very much aware of the fact that they were taking part in an experiment, they didn't know what the specific research questions being addressed with each dish were. For the main course, a piece of Loch Etive salmon, the service staff had laid light canteen cutlery at half of the tables, and heavy, expensive cutlery at the rest.* But still, we weren't asking people about the cutlery; we were only asking them about the food.

The results were unequivocal. Those eating with heavier cutlery thought that their food had been plated more artistically. And crucially, they were willing to pay significantly more for it than those eating the same food, on the same day, in the same dining room, but who just so happened to be holding lighter cutlery instead. So, it really is that straightforward: Adding weight to your guests' hands will most likely make them think that you are a better chef! With that in mind, why not reach into the cutlery drawer right now and feel up some of your own. Are you sure that you are creating the right impression? But you don't want to take things *too* far: I have heard rumours of one restaurant where the diners started to complain because the cutlery was just too heavy to lift comfortably.†

Furry cutlery, anyone?

Periodically, we have lab dinner parties at my house here in Oxford. On one such occasion, Charles Michel, the then chef-in-residence,

---

* The eagle-eyed reader may have noticed that *both* the weight and the quality of the cutlery were varied. However, the fact of the matter is that it is really hard to get heavy poor-quality cutlery. And even if we had found some, there was no way anyone was going to splash out on seventy sets of it just so we could run the study. As is so often the case, then, we had to work within the confines of what was on hand – in this case, canteen versus restaurant cutlery.

† There was a great Pugh cartoon in the *Daily Mail* based on our heavy cutlery findings. It showed a couple dining at the table, with an older women slaving away in the kitchen. Underneath, the punchline read: 'I've changed the cutlery twice and your mother's cooking still tastes horrible.'

Figure 5.4. *Left:* Furry cutlery *à la* Charles Michel. Best served with rabbit; *Right:* Meret Oppenheim's 1936 *Object*, a fur-covered cup, saucer and spoon. How would you feel about putting the cup to your lip to drink? This art piece was deeply subversive at the time, given its sexual connotations. There is just something about fur that you don't want coming too close to your lips. One can only imagine what Freud would have to say on the matter!★

decided to get rabbit from the market and prepare a stew. And it was delicious, I can assure you. But the most memorable thing about the evening, and about that dish in particular, was what the Paul Bocuse-cookery-school-trained chef had done to my wife's cutlery. He had asked at the butcher's for the cleaned rabbit pelts, the stuff that normally just gets thrown away. And in a stroke of *genius* he wrapped them around the handles of the spoon (see Figure 5.4, *left*). In an instant, the cutlery became a truly multisensory dining implement – F. T. Marinetti would have been proud! Sitting around the dining table, we all tentatively held the soft, furry skin in our 'paws', the faint

★ Now, when the chef was working on a short placement at The Fat Duck, I went to dine, and the kitchens came out with something really interesting. Take a tomato and boil to loosen the skin. Try to remove pieces of skin that are as large as possible. Then take the skin and press it in strips on to the lip of a glass, then serve in it something like a gazpacho soup. What you get, as a diner, is a most peculiar sensation, a little animalistic, especially when the inverted tomato skin comes into contact with your lower lip.

aroma of the animal emanating from our hands. There was no doubt – straight away everyone had a much greater awareness of where our dinner had really come from (see Figure 5.4, right, for another famous example).

Imagine my surprise, then, when just a few months later the overweight furry white spoon shown in Figure 5.5 greeted me when I got to the last course of The Fat Duck's revamped tasting menu. Now, I am not sure that this spoon was necessarily the ideal choice. The dish itself is all white, light and airy. So I could imagine a surprisingly light spoon working well here. Instead, what you get is a spoon that is noticeably heavier than you thought it was going to be. (The distinctive baby-powder aroma of heliotropin originates from the handle.) Though perhaps that is the idea: to create even more of a contrast with what you might have expected that you were going to feel. Diners may come away from the experience thinking a little more carefully about the weight of the cutlery, and the influence that it may have on their experience of food.

However, while you want the cutlery to enhance the taste of the food, you don't necessarily want it to become the centre of attention, to distract the diner from whatever they are eating. There is, I think,

Figure 5.5. 'Counting Sheep', the final dish on the tasting menu at The Fat Duck. If you saw a spoon like this, just how heavy would you expect it to be? (It is actually much heavier than that, believe me.)

a very real danger of this happening with the 'Counting Sheep' dish. At least there would be if it weren't for the fact that your dessert is actually spinning on a pillow that is magically floating in mid-air! Yes, really: magnetic levitation, the likes of which you have probably never seen (unless, that is, you have stopped in at the Artesian bar at The Langham Hotel in London, where a cocktail with a balloon floating over it has been on the menu for a couple of years now). This, then, is the aim of much of the research from my lab: to figure out how to use the latest insights from gastrophysics to create better, more memorable tasting experiences.

Eating with your hands

Have you ever thought about the fact that the hamburger, one of the world's most popular foods, is usually eaten with the hands? I would go so far as to say that it tastes better when grasped between one's thumbs and forefingers than when eaten daintily with a knife and fork from a plate or plank. The same goes for fish and chips straight from the newspaper at the seaside too (at least it used to, before someone banned newspaper wrapping as unhygienic). Now, I am the first to admit that there is a lot more going on in the latter case to explain the pleasure of the experience than merely using one's hands. Still, think about it carefully, and it is surprising how many foods really do seem to taste better when eaten like this. No wonder, then, that popular US chef Zachary Pelaccio titled one of his cookbooks *Eat With Your Hands*. He was very much 'on trend'.

And it is not just fast food that people are eating with their fingers nowadays; it is also haute cuisine. For, as we saw at the beginning of this chapter, a growing number of top Michelin-starred restaurants have been incorporating dishes that are to be eaten without cutlery, or else with totally new forms of cutlery.* Interestingly, though – and

---

* Going the whole hog is probably doomed to failure, at least in the context of fine dining. Restaurants like Il Giambellino in Milan, where the diners were not provided with any cutlery at all, don't tend to last long, especially not in the capital of risotto!

I am still trying to figure out the reason(s) behind this – finger food tends to make its appearance at the start of the meal rather than later on. If any of you have any ideas about quite why this should be so, please do let me know.

Many people write to me saying that, for them, food really does taste better when eaten with the hands. This seems to be especially true for those from India, say, who have grown up using their fingers for this purpose. A number of them report that food just appears to lose its taste whenever they have to eat using cutlery. The following, from the Indian narrator in Yann Martel's book *The Life of Pi*, illustrates the point: 'The first time I went to an Indian restaurant in Canada I used my fingers. The waiter looked at me critically and said, "Fresh off the boat, are you?" I blanched. My fingers, which a second before had been taste buds savouring the food a little ahead of my mouth, became dirty under his gaze. They froze like criminals caught in the act. I didn't dare lick them. I wiped them guiltily on my napkin. He had no idea how deeply those words wounded me. They were like nails being driven into my flesh. I picked up the knife and fork. I had hardly ever used such instruments. My hands trembled. My sambar lost its taste.'[4]

I have always wanted to compare what people say when eating and rating a range of foods with either their fingers or with a knife and fork, or even chopsticks. Of course, the answer is likely to depend on the food being served and on the context, not to mention on the individual diner and what they are used to, or have grown up with. Nevertheless, there are some intriguing results already out there showing that the food we feel in the hands really does influence our perception in the mouth. For instance, my colleague Michael Barnett-Cowan conducted a study in Canada in which he somehow managed to glue together two half-pretzels with the opposing ends sometimes having the same texture, either both fresh or both stale, and sometimes having different textures. Just imagine the situation: you might be holding a stale pretzel in your hand but munching on a soft one, or vice versa. The results revealed that the feel of the food in the hand really did influence what people had to say about their in-mouth experience.

Once again, this is a simple change that any one of us could try out the next time that we invite some friends round to eat: withhold the cutlery. A few of you readers may be worrying what Debrett's *Guide to Etiquette* would have to say on the subject. Good news: the 2012 version of the guide finally acknowledged that finger food was acceptable in polite society, at least for certain foods such as pizza, calzone and ice-cream cones. Whatever you do, though, be sure not to lick your fingers afterwards!

And finally, eating with the hands can also be a good idea on a first date, at least if you believe the results of a survey of 2,000 people reported in the papers recently. Men apparently find women making a mess while eating with their hands a big turn-on. So now you know! (And if you are a man hoping to make a good first impression on the ladies, the top tip was to make sure not to order a salad for your main course.)

How much would you enjoy food moving in your mouth?

My colleague Sam Bompas describes a dinner he attended in Korea where the tentacles of a live squid were cut directly on to his plate. His hosts earnestly recommended that Sam chew vigorously in order to prevent the still-squirming suckers from sticking to his throat on the way down! Disgusting, right? So it is ironic, then, that while our *visual* attention is undoubtedly drawn to food in motion (see the 'Sight' chapter), once that food enters the oral cavity, movement is the last thing we want. This deep-seated aversion to things wiggling around in our mouths (and worse still, in our throats) was presumably partly what caused such a stir when Noma served live ants a few years back. (The entomophagy angle not helping much here either.)

It was also the thought of movement in the mouth that made everyone so squeamish when I was at school in Canada for a year, a long, long time ago. The prefects there challenged themselves to eat a live goldfish fresh from the bowl every time the home ice-hockey team scored (which was, thankfully, pretty rare).

Evolutionarily speaking, this dislike of movement in the mouth is probably an old mechanism that helped our ancestors to avoid the risk of choking.* That said, note how the language of menus often tries to give the impression that there is still life in whatever is being served. There's a great line in Steve Coogan and Rob Brydon's TV series *The Trip* where, after a dish is introduced as 'resting' by the waiter, Rob Brydon points out: 'Rather optimistic to say they're resting. Their days of resting have been and gone. They are dead.'[5] True, but 'dead' just feels like a word that should never appear on a menu.

More generally, the texture of food in the mouth (even when it is not moving) seems to be a particularly strong driver of our food likes and dislikes. So, for example, many Asian consumers find the texture of rice pudding to be more off-putting than its taste (or flavour). By contrast, for the Westerner breakfasting in Japan, fermented black natto has a texture and consistency that won't soon be forgotten. And take the oyster – it's this shellfish's slippery, slimy texture, not its taste or flavour, that people typically find so objectionable, many agreeing with the late British food critic A. A. Gill's memorable description of them as 'sea-snot on a half-shell'.

Of course, the textural (oral-somatosensory) properties can also constitute a key part of what we find so pleasing about the foods that we love. Indeed, a number of researchers have argued that this is a key part of the appeal of chocolate, one of the few foods to melt at mouth temperature. (Try eating a very cold versus a warm piece of chocolate to experience this difference.) Texture, then, plays a crucial role in determining our perception of a food's quality, its acceptability and ultimately our food and beverage preferences. Just think about it: comfort foods typically have a soft texture (e.g., mashed potatoes, apple sauce and many puddings). In fact, it has been argued that foods having this texture tend to be thought of as both comforting and nurturing. By contrast, many snack foods are crispy, like chips and pretzels. Texture contrast is something that many chefs and food developers work with and, more generally, it is known to be something

---

* Even food that moves on the plate can itself be both hypnotic and disturbing. The bonito flakes served in Japanese restaurants sometimes do this, not to mention the squirming of the live abalone and hagfish that one sometimes sees at Korean BBQs.

that consumers value in food. As Barb Stuckey puts it in her book *Taste What You're Missing*: 'Good chefs go to great lengths to add texture contrast to their plates, utilizing four different approaches: within a meal, on the plate, within a complex food, and within a simple food.'[6]

What's so special about bowl food?

Hard though it is to believe, five – yes *five* – books were published on the subject of bowl food in 2016. Why, you have to ask yourself, should serving food in a bowl matter? Well, apparently the main appeal is that it makes everything taste better. Even Gwyneth Paltrow thinks so. So it must be true. Of course, it isn't just about serving the same old food in a new receptacle. Part of the appeal to the bowl foodies is filling their receptacles with foods that are wholesome, nutritious and, well, filling.

Serving hot food in a bowl allows, maybe even encourages, the diner to take a hearty sniff of the steaming contents. Most of us are less likely to do this if exactly the same food is served on a plate when it is placed on the table in front of us, say. And as we saw earlier, anything that enhances the olfactory hit associated with a dish is likely to lead to improved flavour perception and possibly also increased feelings of satiety. Holding the bowl in your hands means that you feel its weight too. And the evidence here shows that the heavier the bowl, the more satiated (i.e., fuller) you will expect to feel. This is a problem, of course, for the food and beverage companies, who are being advised by government to make their packaging lighter, especially those trying to promote a filling snack (a yoghurt, say). Time and again in our research, we find that adding weight to a soft-drink can, to a box of chocolates or to a carton of yoghurt leads people to rate the product, no matter what it is, more highly.

Holding a bowl, you feel the warmth of the contents, and possibly the texture and the reassuring roundness of the underside of the bowl. Note here that the texture of plateware has also been shown to modify people's experience of food. In one of our recent studies, for example, we found that people rated ginger biscuits as tasting significantly more spicy when served from a rough plate than from a more traditional smooth plate. Holding a warm cup or bowl in your hand can even make those around you appear a little friendlier too. And, as if all that wasn't enough, serving food from bowls without rims can trick our brains into thinking that there is more than when exactly the same amount is served from a bowl with a wide rim. It is supposedly more photogenic too. Ultimately, then, from a gastrophysics perspective, bowl food is indeed likely to work especially well for those looking for a filling, healthy meal.

Affective ventriloquism

But why exactly should what we feel have such an effect on our taste experiences, especially when what we are interacting with is not the food itself? One possible answer relates to the notion of 'affective ventriloquism'. My colleague Alberto Gallace and I noted a few years ago that people appear to transfer the affective response generated by whatever they touch to what they think about the food or drink itself. That is, we find it hard to maintain separate impressions of the food or drink on the one hand, and of the cutlery, glassware or plateware on the other. Instead, what we think about one can all too easily bias our judgements about the other.

Given that we consume as much as a third of our food and drink direct from the packaging, it should come as no surprise to hear that product designers and marketers are interested in optimizing the feel of their product packaging. In fact, this may be how most of us will be exposed to the whole new world of tactile design. In some cases, the aim is to prime or convey notions of fruitiness, by treating the packaging surface to give it the same feel. The good old Jif lemon juice container is a classic example. In this case, the product itself imitates the size, colour and even feel of a lemon (see Figure 5.6).

However, my absolute favourite comes from the high-end Japanese designer Naoto Fukasawa, a fabulous range of hyper-realistic packaging prototypes, e.g., a drinks container that perfectly captures the experience of touching the fruit – which makes the Jif version of a

Figure 5.6. From left to right: Granini glass bottle; hyper-realistic juice drink packaging by Naoto Fukasawa; and Jif lemon juice container. These examples of multisensory packaging probably enhance the consumption experience by mimicking the feel of the fruit they contain.

lemon feel cheap by comparison. Amazingly, the designer has perfectly rendered the surface of a banana, of a strawberry and, most impressive of all, the hairy skin of a kiwi fruit.

It is more than fifteen years ago now since we started working with Unilever on exactly this topic. Our idea then was to try to enhance the fruity notes in Lipton peach-flavoured iced tea by treating the packaging to give it something of a furry-peach feel. At that time, the solution was just too expensive to be practicable, but – good news for food and beverage companies – giving packaging surfaces a unique and realistic feel is now becoming much cheaper. And I am more convinced than ever, given the research evidence of the intervening years, that getting the feel in the hand right is an important way in which to improve the consumer's experience of food and drink in the future.

No matter whether it be cutlery, glassware, plates or bowls, the gastrophysics approach is providing the evidence and insights to support the creative tactile designs that are now emerging for the dining table. The most extreme and intriguing examples will come from designers, modernist chefs and molecular mixologists. However, my prediction is that the majority of us will be exposed to this new approach through food and beverage packaging – everything from the textured paint on a Heineken can (brought out in 2010, the special cans were intended to have a 'signature' feel) to the silky surface of some high-end boxes of chocolates.

6. The Atmospheric Meal

When we imagine an eating experience, we can't ignore the setting. If you are far from home, in some foreign city, deciding where to eat, don't you always end up gravitating towards the restaurant that has a buzz, a vibe, in other words, the one that has 'atmosphere'? And don't we all mostly stay away from those places where there aren't any diners – you know, restaurants that look stone dead – no matter how highly they come recommended?

Can the atmosphere in a restaurant determine how much we eat, not to mention how much we spend? Many restaurateurs certainly believe so. As the owner of Pier Four in Boston said, back in 1965 when it was one of North America's most successful restaurants: 'If it weren't for the atmosphere, I couldn't do nearly the business I do.'[1] But beyond its effect on table turnover, not to mention maximizing the bottom line, can you really enhance the perceived taste and/or enjoyment of a meal just by picking out the right sort of background music? Does the *same* food actually taste *different* when the atmosphere, or environment, in which it is served changes? As I will show you in this chapter, emerging gastrophysics research shows that the answer to these questions is very often yes.

Environmental attributes, from the music through to the lighting, and from the ambient scent through to the feel of the chair you are sitting on, can – and in many cases do – influence the dining experience. Marketers have long been aware of the profound influence of the environment. The famous North American marketer Philip Kotler, for instance, in his seminal early paper on 'atmospherics', emphasized how a key part of the total product offering is the atmosphere in which a product or service is presented, which is itself multisensory. He drew a distinction between the *tangible product* and the *total product*. He is, I think, worth quoting at length given how influential this work has been: 'The *tangible product* – a pair of shoes, a

refrigerator, a haircut, or a meal – is only a small part of the total consumption package. Buyers respond to the *total product*. [. . .] One of the
more significant features of the total product is the *place* where it is
bought or consumed. In some cases, the place, more specifically, the
atmosphere of the place, is more influential than the product itself in
the purchase decision. In some cases, the atmosphere is the primary
product.'[2]

To date, the majority of the research on atmospherics has tended to
focus on music – the easiest aspect of the environment to change. So,
to begin with, let's take a look at the evidence concerning the effect of
background music on our dining behaviour.

Moving to the beat

Do you think that you would eat and drink more rapidly if the tempo
or loudness of the music in a restaurant increased? Would you end up
spending more if they were to play classical rather than top-forty
music, say? And would you be more likely to choose something
French if accordion music was playing in the background? Sounds
unlikely, right? Yet in one of the most impressive demonstrations of
background music's impact on consumer purchasing behaviour, this
is exactly what was found. In particular, researchers alternated the
type of music in the wine section of a British supermarket. When
French music was played, the majority of people bought French wine;
however, when distinctively German (Bierkeller) music was played
instead, the majority of wines sold were German. The numbers have
to be seen to be believed (see Figure 6.1).

Most people, when told about such results, are convinced that
they wouldn't be so easily influenced. So too, in fact, were the customers who were questioned on leaving the tills in the study itself, the
majority of whom resolutely denied that the music playing in the
background had swayed their purchasing decisions that day. They
confidently asserted that they had always intended to buy French
wine, as the accordion music played in the background. However, the
sales figures tell a very different story. Given results like these, you

Background Music

	French accordian music	German Bierkeller music
Bottles of French wine sold	**40** (77%)	**12** (23%)
Bottles of German wine sold	**8** (27%)	**22** (73%)

Figure 6.1. Number (and percentage in brackets) of bottles of French versus German wine sold as a function of the background music in one of the most oft-cited marketing studies on the impact of ambient music on people's behaviour.

can probably better understand why gastrophysicists are often so sceptical of people's subjective reports. Better to look at what people do, rather than merely relying on what they say.

Do you think that your food preferences would be affected by a restaurant deciding to change its decor? Well, one study that went some way towards answering this question was conducted at the Grill Room at Bournemouth University, in the UK, back in the early 1990s. Researchers wanted to know whether they could alter the perceived ethnicity of a range of Italian dishes without changing the food offering. To that end, a selection of Italian and British foods was offered up over four days. On the first two days, the restaurant was decorated as normal (e.g., with white tablecloths and the walls and ceiling unadorned). For the other two days, the restaurant was given an Italian feel: Italian flags and posters were mounted on the walls and ceiling, the tables were covered with red and white chequered tablecloths. Oh, and a wine bottle was placed on each table for good measure.

The diners (138 of them, to be precise) were invited to complete a questionnaire once they had finished their meal. They were asked how ethnic their meal had been, as well as how acceptable they found the food overall. Giving the restaurant an Italian theme resulted in diners choosing more pasta and Italian desserts such as ice cream and zabaglione, and significantly fewer fish dishes. Adding an Italian feel also resulted in the pasta items being rated as more authentic. The perceived ethnicity of the meal as a whole went up too, with 76% of the diners describing the meal as Italian, as compared to only 37% in the baseline condition. Such results illustrate how what people think about a meal can be influenced by changing nothing more than the visual attributes of the environment in which that dish happens to be served. And given what we have already seen, had Italian music been played too, who knows how much more pronounced the multisensory atmospheric effects would have been?

So at home, you could perhaps make your pizza and pasta taste more authentic by playing a bit of Italian opera. The film director Francis Ford Coppola, for one, insisted on 'musical accompaniments matched to menus – accordion players for an Italian *pranzo*, mariachi for Mexican *comida*' whenever he was filming.[3]

The question remains, what might work best, practically speaking, in terms of music to accompany your takeaway pizza? Well, you'll be glad to hear that here at the Crossmodal Research Laboratory we have done the relevant study in order to determine just that (even if Italian music might make Italian food seem more authentic, it still might not give rise to the best experience for the diner). In a recent project conducted on behalf of Just Eat (an online food-ordering company, like Seamless in the US), we asked more than 700 consumers which of 20 musical tracks worked best with the 5 most common types of takeaway food in the UK: Italian, Indian, Thai, Chinese and sushi. The music spanned several genres, from R'n'B and hip-hop, pop and rock, through to classical and jazz. Pavarotti's 'Nessun Dorma' came out as the top match for takeaway Italian food. Across the board, Nina Simone's 'Feeling Good' and Frank Sinatra's 'One for My Baby' were always ranked in the top three tracks, regardless of the type of takeaway that our participants were evaluating. So they too would seem to be a safe bet for anyone without a compendious music collection. But the really big surprise was that Justin Bieber's 'Baby' came pretty much bottom. So that one is a big no-no – you have been warned. (Sorry, all you Beliebers out there . . . but you can't argue with the data!) As to why the results turned out this way, well, that is a question that we are still working to answer.

Playing classical music in the background tends to result in people splashing out more. This turns out to be true no matter whether one is looking at diners' willingness to pay for the food served in a student cafeteria or customers' actual spending behaviour in a restaurant setting. In fact, a 10% increase in the average bill is not unheard of. For instance, in one study conducted by Professor Adrian North, at Softley's restaurant in Market Bosworth, Leicestershire, the diners spent £2 a head more, on average, when classical music was playing in the background rather than pop. In another study, the customers in a wine store were shown to spend more when classical music rather than top-forty music was played.

Background music can impact how pleasant we rate the food itself. No prize for guessing that the more discomfiting the music, the less time people spend in a given environment, while the more they like the music, the longer they stay. And, generally speaking, the more you like the music, or the environment, the more you like the food and drink. That said, I always suggest to clients that they should conduct their own research in order to know what kind of musical backdrop will work best for them. There may well be some significant cross-cultural differences here in terms of the appropriateness of music (and/or conversation) at mealtimes too. In Korea and Japan, for instance, it is far more common to eat a restaurant meal in silence without people talking much to one another, and with no music in the background.

That said, it is important to bear in mind the congruency between the restaurant concept, the clientele and the type of music that is played. It is hard to believe that classical music would necessarily be the right choice in one of those grungy burger bars that are popping up all over the place these days. It just wouldn't seem appropriate, would it? But for a high-end hotel restaurant, where the mark-up on the French wine is much higher than for any of the other stuff that they have in the cellars, the implications are clear. What we can infer here is that classical music is more likely to prime notions of class, and/or that those who are drawn to classical music may, on average, be a little more affluent.

Next, let's take a look at tempo, i.e., the number of beats per minute (bpm), and loudness. Is the speed at which you eat or drink influenced by the speed of the music playing in the background? By now, I am sure you can guess what the answer to this question will be. And indeed, playing faster music has been shown, across a number of studies, to result in people eating and drinking more rapidly. In what is perhaps the classic study in this area, this time conducted by R. E. Milliman, a North American professor of marketing, back in 1986, the tempo of the music playing in a medium-sized restaurant was manipulated. The 1,400 North American diners whose behaviour was assessed ate much more quickly when fast (as compared to slow) instrumental music was played. When slow music was played, diners spent more than 10 minutes longer eating, bringing the total duration of their restaurant stay up to almost an hour. Although the musical tempo had no effect on how much money people spent on their food, there was a marked difference on the final bar tab, with those exposed to the slow music spending around a third more! Slowing the music down increased the restaurant's gross margin by almost 15%, no doubt a sensible idea at quiet times. However, when the customers are queuing out of the door, the restaurateur might be better off playing some fast-paced music instead.

But would a restaurant chain really go to such lengths to control the flow of customers? Absolutely! Just listen to Chris Golub, the man responsible for selecting the music that plays in all 1,500 Chipotle restaurants in the US: 'The lunch and dinner rush have songs with higher bpms because they need to keep the customers moving.'[4] In fact, Golub is often to be found sitting in his local New York branch of Chipotle, observing people's behaviour in response to the different music that he is thinking of adding to the playlist. Then, depending on the customers' responses, he will fine-tune it both for tempo and style before it is beamed out to branches across the land. All that is missing here is the statistical analysis and this would be fully-fledged gastrophysics research!

Ultimately, of course, restaurateurs and bar managers are primarily interested in increasing their profits. The Hard Rock Café chain, for instance, plays loud music in its venues because of the positive effect on sales. Just take the following, from a piece that appeared in *The New York Times*: '[T]he Hard Rock Café had the practice down to a science, ever since its founders realized that by playing loud, fast music, patrons talked less, consumed more and left quickly, a technique documented in the International Directory of Company Histories.'[5] And according to another report: 'When music in a bar gets 22 per cent louder, patrons drink 26 per cent faster.'[6] This shines a whole new light on why so many restaurants and bars are getting louder than ever before. Put simply, it makes us spend more!

That said, much of the underpinning research was conducted years ago, in a different time and place; the results may no longer apply. My recommendation to the restaurateur is just to be aware of the importance of the atmosphere to the food offering you provide. This should at the very least help you to avoid the situation whereby the chef who obviously cares passionately about their food allows the front-of-house manager to put their iPod on random shuffle. You know the sort of thing: you suddenly find yourself listening to Frank Sinatra singing 'Jingle Bells' in the middle of July when eating in a Thai restaurant, say! This really shouldn't happen. But I think we all know that it does, and more often than it ought to. If you have the opportunity to experiment, why not try French music this week and American Rock next, or classical music today and the top-forty tomorrow, and see for yourself the impact on what people say (or, more importantly, on sales). Gastrophysics research provides a number of suggestions as to what might happen, but you will have to check for yourself to be sure of getting the effect you want. Broadly speaking, though, when the atmosphere matches (or is congruent with) the food offering, people will tend to enjoy the experience more.

Do you care about being comfortable?

Have you ever wondered why most trendy coffee shops have such hard and uncomfortable seating? Well, put simply, they just don't want you to linger. I know of a number of baristas who deliberately chose hard, uncomfortable furniture in order to discourage their customers from loitering and hogging the tables all day long. You don't need a gastrophysicist to tell you that the less comfortable the chair, the shorter your stay will be. McDonald's has been doing this for years; as one commentator put it: 'The rule written into the design of the seats [in McDonald's] is that 10 minutes is the appropriate length of one's stay before they become uncomfortable.'[7] At the high end, though, where the duration of the customer's stay isn't an issue, restaurants are increasingly thinking about how to augment the feel of the space in which their food is served. A few innovative chefs, such as Joshua Skenes, chef-owner of Saison in San Francisco, have started to play with giving their restaurants a distinctive feel. According to the chef: 'You need great food, great service, great wine, great comfort. And comfort means everything. It means the materials you touch, the plates, the whole idea that the silverware was the right weight. We put throws on the back of the chairs.'[8] Take a look at Figure 6.2; it seems Noma in Copenhagen are doing much the same thing.

Would you rather sit at a round or a square table? Generally speaking, people prefer round (or curvilinear) to angular forms, a preference that extends all the way from everyday objects through to architectural spaces and even furniture. Some evolutionary psychologists put

Figure 6.2. A chair with texture at Noma in Copenhagen.

this seemingly ubiquitous preference down to angular forms being associated with danger (think sharp and dangerous weapons). Of course, due to practical constraints, the majority of traditional restaurant floor plans are angular. But a square space can be the frame for much rounder forms, in the decoration or the furniture.

In one recent study, a group of North American university students was shown pictures of interior environments containing either angular or rounded furniture. The results highlighted a preference for rounder furniture, with the latter tending to elicit greater feelings of pleasure as well. Interestingly, in this case, the participants reported a greater desire to approach the curvilinear rather than the rectilinear furniture. As one of the participants said: '[R]ounded furniture seems to give off that calming feeling.' Round tables can be used to help make the interior of a restaurant appear more welcoming. But they also reduce capacity – presumably why many restaurant consultants recommend a mixture of round and angular tables, aiming for a balance between approachability and profitability.

How would you like to dine in a white cube?

There are some traditional establishments where no attempt has been made to augment the atmosphere whatsoever: Just think of those temples to haute cuisine with their unadorned white walls, diners sitting before a starched white tablecloth (or, as is currently fashionable, just a starched napkin), eating in hushed and respectful silence. No one could claim that such venues were trying to distract their diners from the food, right? The idea of releasing ambient fragrance or changing the temperature of the space to match the dishes being served would, one imagines, be complete anathema to such traditional restaurateurs. I guess that there will always be a place for such austere dining rooms. But my sense is that it is hard to make this kind of approach seem contemporary or exciting, at least in the current climate. More often than not, such venues are being replaced (in the San Pellegrino listing of the world's fifty best restaurants, say) by the more experiential dining concepts.

What is more, bear in mind that by removing atmospheric cues one is still making a statement. Here, I am reminded of a quote from one commentator: 'The modern restaurant is an experience in codes. The architecture, the foods served, and even the customers are codes built up to the total consumable image. Restaurants then do not just serve food. They serve an experience.'⁹ So while the decor may be minimalist, the atmosphere is very definitely never 'neutral'. Make no mistake, the dishes served in the 'white cube' environment will be rated differently by diners than in any other environment that one might care to choose. Based on the evidence, the food would probably be rated as better in quality and more expensive, though perhaps less memorable. The key point is that there is always an atmosphere wherever food is served and consumed.

The same also goes for those healthy, natural, organic stores and restaurants, the places with the baskets of fresh produce on display as you walk in (many of Jamie Oliver's restaurants fall into this category). Make no mistake about it, this kind of atmosphere is itself priming notions of healthy and natural in the mind of the diner. It may look casual, but it most certainly isn't – the creation of the display is itself artifice. Often, a great deal of thought has gone into constructing that 'natural' environment to be just so. That is part of the conceit; in fact, I bet it affects the experience just as much as in other restaurants where the atmosphere changes on a course-by-course basis. They are still creating an impression, an expectation that will colour the whole encounter between the diner and their food and drink.

Over the years, some restaurants have really gone overboard in terms of delivering a multisensory atmosphere. One of the most famous early examples here is the Tonga Room & Hurricane Bar, which opened down in the basement of the Fairmont hotel in San Francisco back in 1945. I still remember visiting as a young graduate student, long before my current interest in multisensory dining had taken root. A spectacular tropical thunderstorm, with simulated thunder and lightning, would unfold every thirty minutes or so during opening hours. Good idea though this was, presenting the same old sound and light show for so many years had taken its toll on the place – it was looking a little tired. Furthermore, it is easy to imagine how the customers might well habituate to the repeating multisensory scene that unfolds.

A little over five decades after The Tonga Room first opened its doors to the public, and on the other side of the pond, as it were, one finds The Rainforest Café. This well-known London restaurant also delivers an experience that tries to stimulate *all* of the customers' senses. Every half an hour or so, the restaurant goes dark while the guests are 'treated' to all the rumbling and flashing of a thunderstorm in the tropics. While The Tonga Room targets a more mature clientele, the latter venue obviously has its sights set on a much younger market (or rather, on the pockets of those who have been charged with looking after them). As the self-styled engineers of the experience economy (see the 'Experiential Meal'), B. J. Pine, II, and J. H. Gilmore, say: 'The mist at the Rainforest Café appeals serially to all five senses. It is first apparent as a sound: Sss-sss-zzz. Then you see the mist rising from the rocks and feel it soft and cool against your skin. Finally, you smell its tropical essence, and you taste (or imagine that you do) its freshness. What you can't be is unaffected by the mist.'[10]

No matter whether the grown-ups like it or not, there can be no doubting how incredibly successful the experience has been amongst the target audience, namely children. For a few years, my nieces were huge fans, though I suspect that they have rather grown out of it now. And what is absolutely clear is just how successful the venture has been, commercially speaking. In other words, atmospherics sells, or at least it does when done well.

There has sometimes been, it has to be said, a lingering suspicion that restaurateurs are interested in the atmosphere only insofar as it differentiates them from the competition and increases their bottom line. Of course, it is all too easy to get a little sniffy about the grubby financial side of things, but who isn't ultimately interested in at least breaking even? As the influential British chef Marco Pierre White once put it: 'Any chef who says he does it for love is a liar. At the end of the day it's all about money. I never thought I would ever think like that but I do now. I don't enjoy it. I don't enjoy having to kill myself six days a week to pay the bank . . . If you've got no money you can't do anything; you're a prisoner of society. At the end of the day it's just another job. It's all sweat and toil and dirt: it's misery.'[11]

One could also mention the popular dine-in-the-dark restaurants here (e.g., Dark Restaurant Nocti Vagus in Berlin) – they clearly fit into the atmospherics framework, with a sensory input removed, rather than added. Nevertheless, going to one of these restaurants is very definitely an experience, but not necessarily one that is centred on great-tasting food.

To summarize, then, the atmosphere affects our food behaviours in a number of ways: everything from influencing where and what we 'choose' to eat, through how long we stay, not to mention what we think of the overall experience (see Figure 6.3). But it is worth noting that we haven't so far addressed the more fundamental question of

Figure 6.3. All of the senses play a role in controlling our behaviour while drinking and dining. The intelligent restaurateur knows how to work with the senses to create the right environment. The scientific approach to multisensory atmosphere design has led to a number of restaurant chains increasing their profitability.

whether changing the environment really does modify what people perceive on the plate or in their glass. This is the kind of question that is of most interest to the gastrophysicist.

Atmospheric tasting

Listen to the chefs and you'll hear conflicting views. When interviewed a couple of years ago, French chef Paul Pairet was quoted as saying that he didn't believe that all the multisensory atmospherics in his restaurant Ultraviolet in Shanghai made any of his dishes taste better. Rather, he thought simply that 'the memory of the dish is stronger'. A worthy enough aim, but is that all there is to it? Ironically, the press report in which Pairet is quoted itself seems to come to a different conclusion. For according to the journalist: 'each dish is accompanied by a carefully choreographed set of sounds, visuals and even scents, all intended to create a specific ambience to enhance the flavors of the meal.'[12] Pairet is not alone in his views, though. Others include French chef Alain Senderens, who once complained about the Michelin man's penchant for fancy fittings. 'I was spending hundreds of thousands of euros a year on the dining room – on flowers, on glasses,' he said, 'but it didn't make the food taste any better.'[13]

In the other camp there are those, like Heston Blumenthal, who have latched on to the fact that the atmospherics really can modify the tasting experience. We first demonstrated this with Heston at the 'Art and the Senses' conference held in Oxford back in 2007. The lucky participants at this event got to eat oysters while listening to the sounds of the sea and to taste bacon-and-egg ice cream to either the sound of sizzling bacon or the clucking of farmyard chickens. We demonstrated that people rate bacon-and-egg ice cream as tasting significantly eggier when listening to the sound of farmyard chickens clucking in the yard, but when we played the sounds of sizzling bacon instead, suddenly the bacon flavour became rather more intense. Changing the atmospheric sounds altered people's perception of the test foods. Playing the sounds of the sea also made the oysters more pleasurable (but no saltier).

In the years since, I have been lucky enough to team up with some of the world's leading drinks brands to conduct various large-scale multisensory tasting events with members of the public. Typically, these events have built on the belief that changing the atmosphere will influence the tasting experience. And rather than manipulating just the sonic environment, we have been working to modify the visual and olfactory aspects of the environment too. Let me share a couple of these experiences with you.

'The Singleton Sensorium'

Typical of this gastrophysics approach was 'The Singleton Sensorium', which took place over three evenings, in the heart of Soho, London, in 2013. My colleagues from Condiment Junkie, a UK-based sound agency, decorated three rooms in an old gun-maker's studio in very different styles. One room aimed to recreate an English summer afternoon, another was designed to prime notions of sweetness, while the third room had a distinctly woody theme. The Condie boys also generated some atmospheric soundscapes to play in the background in each room. Take the sweet room. It was decorated in a pinky-red hue, chosen because that is the colour that most people generally associate with sweetness. There was nothing angular in the room; everything was round – the pouf, the table, even the floor plan and the window frames. Why? Well, because our research had shown that people associate rounder shapes with sweetness. There was also the sweet-smelling but non-food-related fragrance and the high-pitched tinkling of what sounded like wind chimes coming from a ceiling-mounted loudspeaker. The latter choice was again based on our laboratory research showing that people associate such sounds with sweetness. So every sensory cue had been chosen on the basis of the latest gastrophysics research to help prime, consciously or otherwise, notions of sweetness on the palate. The first room, by contrast, had been designed to prime grassiness on the nose. The final 'woody' room was meant to prime a textured finish, or aftertaste, in the mouth.

Over three evenings, nearly 500 people were escorted in groups of

10 to 15 through an experience lasting no more than 15 minutes. At the outset, everyone was given a glass of whisky, a scorecard and a pencil. They filled in one section of the scorecard while standing in each room. The punters were asked about the grassiness of the whisky on the nose, the sweetness of its taste, and the woody aftertaste. They indicated how much they liked the whisky, and what they thought of the decoration in the room that they were standing in. I was one of the tour guides, and let me tell you, it was an exhausting experience. It was the first time that anything like this had been tried on this scale. Would the experiment work as planned, or would people simply walk away saying that the whisky obviously tasted the same in all three rooms because it was, after all, the same whisky?

It was a huge relief to find, once the results were analysed, that, as a group, people rated the grassiness of the nose of the whisky as significantly more intense in the grassy room. Meanwhile, the second room brought out the sweetness on the palate (as expected), and the final (woody) room really did accentuate the textured finish of the whisky. As a psychologist, one always worries about so-called 'experimenter expectancy effects', i.e., that your subjects may say what they think you want to hear, rather than tell you what they actually experienced or thought. In fact, at the end of the 'Sensorium', one or two people did approach me and say something of this sort: 'We knew what you were up to. You wanted us to say the whisky tasted grassier in the green room, right? So we did the opposite!'

Note, though, that even these truculent individuals were not *unaffected* by the multisensory environment (at least in a manner of speaking). And crucially, the group analysis revealed that such individuals were clearly in the minority. What is more, people enjoyed the whisky most in the woody environment. So, manipulating the multisensory atmosphere in this scientifically inspired manner really did affect what people had to say about the drink in their hand. Depending on the room, the change in people's ratings of the nose, taste and finish of the whisky were in the order of 10 to 20%.

Would whisky experts have been equally affected by 'The Singleton Sensorium'? It is difficult to say for sure. However, it is worth noting that neither the whisky expert nor, for that matter, the wine aficionado can necessarily do all of the things that they think (or say) that they can when it comes to blind-tasting. What is perhaps more relevant is that the experience was powerful enough for a number of chefs, restaurateurs and designers to go away and change the way in which they delivered some of their food and beverage offerings. For example, staff at one famous restaurant in the Lake District, in north-west England, started serving guests whisky from a wooden tray, matching the environment in which they themselves had most enjoyed the drink while taking part in the event.

'The Colour Lab'

What do you think would be the best colours to bring out the fruitiness and freshness of a wine? And could you achieve the same effect by playing sweet or sour music (sour music tends to be dissonant, high-pitched, rough, sharp and staccato)? We set out to answer these questions in what may well have been the biggest ever tasting event of its kind, known as 'The Colour Lab'. More than 3,000 people were tested over an unseasonally warm May Bank Holiday weekend on the banks of the River Thames in London, as part of 'The Streets of Spain' festival. Each person was given a glass of Spanish Rioja in a black glass. They had to rate the wine first under regular white lighting (to get a baseline measure), then under red illumination, then again in a green environment with 'sour' music. Finally, they tasted the wine under red lighting again, but this time accompanied by 'sweet' music. Once again, a 15 to 20% change in people's ratings was observed on switching from one audiovisual atmospheric combination to another. The red lights and sweeter music (consonant, high-pitched, neither rough nor sharp but smooth and flowing) were found to accentuate the fruitiness in the wine, while the green colour and sourer music brought out its fresher notes instead.

While previous gastrophysics research had already demonstrated (albeit on a much smaller scale) that changing the colour of the light bulbs or the music playing in the background can change what people say about the wine they are tasting, we were the first to combine the senses in a multisensorially congruent manner. We were looking for what some have termed a 'superadditive' effect. Put simply, this is the idea that the various atmospheric cues might combine to deliver a multisensory effect that was bigger than the sum of its parts (i.e., greater than what you might expect merely from adding the effect of light and sound when presented individually). And as we had hoped, the sonic seasoning – sweet music with red lighting and sour music with the green lighting – did indeed enhance the effect of the lighting on the taste of the wine.

One of the results (or deliverables) of such multisensory events is the statistical evidence that the environment influences people's perception. On occasion, the results may also reveal the relative importance of the senses to that experience. Often, though, what is much more powerful, at least in terms of convincing people, is the change in what they themselves felt. In fact, when we trialled 'The Colour Lab' on the wine makers from Campo Viejo, they were so impressed that they left saying that they would be redesigning their own cellar-door experience as soon as they got back to Spain. Furthermore, one of the wine writers with whom I work, who was, by his own admission, initially sceptical, now uses changes in the ambient lighting as a party trick when leading informal wine tastings. So next time you open a bottle of wine at home and find that it is not quite to your taste, why not just try changing the music and/or the lighting first before you reach for a different bottle? Sometimes, it really is that simple (assuming that the wine doesn't have any obvious flaws). Nowadays you can buy remote-control colour-changing light bulbs for virtually nothing online. So there really is no excuse, is there?

If you are wondering what counts as sour music, you could try Nils Økland's 'Horisont'. For something sweet, look for tracks with lots of tinkling, high-pitched piano notes. I often use something like 'Poules et Coqs' from Camille Saint-Saën's *Carnival of the Animals* or tracks 6 and 7 from Mike Oldfield's *Tubular Bells* (1973). Choose something like *Carmina Burana* by Carl Orff or 'Nessun Dorma' from the third act of Puccini's *Turandot* if you want to bring out the depth in a red wine, say a Malbec.

These lighting manipulations work well enough when the wine is served in a black tasting glass (as in 'The Colour Lab'). I could well imagine that the effects would be even more pronounced if the wine were to be served in a clear glass, where the colour of the wine itself may also change as a function of the ambient lighting. You do need to be a little careful here, though, because if too dramatic a change in the lighting is introduced at mealtimes, it may actually change the visual appearance of the food itself. As one commentator put it: 'A red light makes everything look red; a green light makes meat look grey and spoiled.'[14]

Of course, different people have different objectives as far as the use of ambient lighting is concerned. While some may wish to bring out the freshness in their wine, others may be wondering whether there are certain colours or, for that matter, types of music, that may help to promote healthy eating, for instance, use of red lighting to bring out sweetness (without the calories). Research shows that the colour of the ambient lighting can influence a diner's appetite. For example, yellow lighting was found to increase people's appetite, whereas red and blue lighting decreased people's motivation to eat. When the colour of the food and of the ambient lighting match, it seems to stimulate appetite, whereas complementary colours suppress it. Also relevant here are the results of a recent study conducted in Sweden, in which it was found that dieting Swedish males eating their breakfasts under blue lighting felt fuller with less food.

Controlling the restaurant environment

So we know what happens in experiential events. But do you think that you would eat less if the lighting and music in a restaurant were softened to create a more relaxed atmosphere? Well, researchers tested the combined impact of changing the lighting and music on the behaviour of diners in Hardee's, a fast food restaurant in Champaign, Illinois. The restaurant in question had two dining areas (ideal for gastrophysics research). In one, the lighting was set at its normal bright level, the colour scheme was also bright and the music playing in the background was loud. The other 'fine dining' environment had a much more relaxed atmosphere: there were pot plants and paintings, window blinds and indirect lighting. Oh, and did I mention the white-tablecloth-covered tables with candles on top, and soft jazz instrumental ballads in the background? Those who ate in the more relaxed side of the restaurant rated their meal as significantly more enjoyable, while at the same time consuming less (their calorie intake was reduced by an average of more than 150 calories, or 18%).

The fact that the environment has such an effect on us could obviously have implications for the restaurateur. In fact, it has been suggested that this is precisely why the Hard Rock Café and Planet Hollywood chains have no windows, which gives them (just like the casinos) greater control over the environmental stimulation their customers are exposed to.

The future of atmospherics

So, how might the atmospheric aspects of dining change, moving forward? As one designer put it recently: 'In the short time I have been in the business of designing restaurants, design has definitely become a major element of the dining experience. The environment and its uniqueness are becoming as important as the food, and designers and owners are becoming more sophisticated in how they use light, colour and materials.'[15] And if you want to see what the future may hold in terms of restaurant design, why not take a look at the Goji Kitchen & Bar at the Marriott Bund hotel in Shanghai. The decor in this futuristic dining space actually changes to give the restaurant one of two different feels, depending on the time of day. This is undoubtedly an expensive solution, but it is also an acknowledgement that decor and atmosphere really do matter. It stands as testament to the importance of the atmospheric component of 'the everything else' to mealtimes. And while it is always difficult to figure out how much to spend on the decor, as soon as you are aware of just how much it influences the dining experience, then there really is no looking back. However, while getting the atmosphere 'right' is undoubtedly a challenge, it is also important to remember that there is no way to avoid its influence, much though one may wish to.

There is an interesting challenge here in terms of how to customize the atmospherics to individual diners, or tables. Currently, most high-end multisensory dining experiences involve either a single-sitting restaurant (think Ultraviolet in Shanghai or Sublimotion in Ibiza), or headphones being brought to the table to accompany a particular dish (the 'Sound of the Sea' at The Fat Duck, for instance). However, I know of restaurateurs out there who are already thinking about whether they can use hyper-directional loudspeakers positioned over the tables in order to deliver a soundscape personalized to whatever the diners happen to be eating and drinking. Crucially, none of the diners would be able to hear what was going on at any of the other tables. This kind of solution is currently prohibitively expensive for all but the most affluent of restaurateurs. Nevertheless, looking ahead, I can well imagine that it will become more common, given the growing emphasis on personalization and customization, together with the falling cost of technology.

In this light (if you'll excuse the pun), there are now multi-coloured LEDs installed over each of the tables at The Fat Duck restaurant, following its recent refurbishment. These bulbs subtly change colour as the diners at a given table progress on a journey from night to day and on to the next evening during the course of their meal. The lighting changes take place at different times on each table. Is this really the future of personalized atmospherics? I suspect that it might be a start.

7.   Social Dining

I don't know about you, but I don't like dining out alone. And so a human-interest story in the papers recently caught my attention. It concerned a nonagenarian British widower, Harry Scott, who had been dining alone at his local McDonald's pretty much every day, sometimes twice a day, for the last three years. The sad truth was that he simply has no one else to share his meals with since his wife died. So, on the occasion of his 93rd birthday, the staff at the branch in Workington, Cumbria, held a party for the old geezer. I have to say, looking at the picture that appeared in the papers at the time, that Harry appeared to be in much better shape than you might have expected.[1]

While this is an isolated case, I nevertheless think that it is indicative of what is happening in society at large. For the fact of the matter is that a growing number of us are dining alone (see Figure 7.1). Nearly half of all meals are now eaten alone, and more than a quarter of us eat by ourselves more often than we eat in company, according to one recent British survey. Worse still, many of us eat our main meal in isolation most days (perhaps eating lunch alone at a desk, or consuming a microwaved meal, or grabbing a bite to eat at a drive-thru). Given that the figures vary by culture and age, why not count up how many meals *you* have eaten by yourself over the last week, say, and see whether or not you buck the trend.

But why, you might ask, should we even care about this change in our eating behaviour, beyond being concerned by the growing isolation that it hints at in society today? And what exactly has the company we keep got to do with the experience of the food we eat, that is, with the topic of this book, namely gastrophysics? Not everybody thinks we should be worried; here's Nell Frizzell, writing in the *Guardian* : 'Like life's other great pleasures, eating alone is something you can do one-handed, lying on your back, in nothing but an old jumper, should you so wish. It isn't lonely, it isn't distasteful, it isn't desperate: it's a celebration of existence. It keeps us alive – as simple as that.'[2] I beg to disagree. As we'll see later, much of the evidence points towards dining solo having a negative impact on people's physical health and mental well-being. A recent meta-analysis of seventeen different studies, involving more than 180,000 adolescents and children, revealed that regularly sharing meals as a family reduced the odds of youngsters being overweight by 12%. It also increased the likelihood of their eating healthy foods by almost 25%. Not only that but as a gastrophysicist I would agree with American psychologist Harry Harlow, who, back in the 1930s, put it thus: '[A] good meal tastes better if we eat it in the company of friends.'[3] Gastrophysics offers a constructive framework in which to look for solutions to the growing problems associated with solo dining.

Figure 7.1. Dining solo – a growing problem in society today.

Why are so many of us eating alone?

Part of the reason is undoubtedly linked to the fact that more of us are *living* alone than ever before, due to people marrying later, divorce rates going up and people living alone for longer. Another important factor here relates to our changing food habits. For one thing, there are simply far fewer meal-based family gatherings than at any point in the recent past. When was the last time you asked anyone over to your house for dinner? According to the results of one recent survey, 78% of Brits virtually never invite their friends over at mealtimes any more. When quizzed about why, the reason that people often give is that they just find it too much of a chore to prepare food from scratch, given their increasingly hectic lifestyles. Indeed, the average time spent preparing a meal has dropped from around an hour back in 1960 down to just 34 minutes today. What all of this means is that one in three of us will now go an entire week without eating dinner with another person.

Is solo dining really so bad?

Dining solo is bad for us in a number of ways. On the one hand, those who eat by themselves tend to engage in poorer food behaviours. In males, for instance, eating and living alone are jointly associated with a higher prevalence of weight problems, by which I mean obesity at one end of the spectrum and being underweight and/or engaging in unhealthy eating behaviours, such as a low fruit and vegetable intake, at the other. Unsurprisingly, those who dine solo are also more likely to feel lonely. Many older individuals who find themselves in hospital or in long-term care also suffer from undernutrition, which is made worse by their often being forced to eat solo. *Anything* that can be done to bring the social element back to dining is likely to help improve the nutritional status of these vulnerable individuals. For instance, a couple of studies conducted in the US demonstrated that elderly hospitalized patients end up consuming significantly more food when they are encouraged to engage in more active, inter-personal behaviour with their care-givers at mealtimes. Those who live by themselves, and hence regularly eat alone, also tend to generate far more food waste than those who live and dine with others, a problem exacerbated by the fact that the portion sizes that are commonly available in the supermarket often do not cater to those living alone. According to a 2013 UK Government survey, people living alone throw away 40% more of their food than those living with others.

Dining with distraction

The decline of the more social aspects of dining is not restricted to the increasing proportion of us who, for whatever reason, eat alone. Technology also has its part to play. How often, after all, do *you* have the TV on at mealtimes? And how often do *you* find yourself with fork, spoon or chopsticks in one hand and your smartphone in the other? Even those of us who are physically sitting together around the dining table are too often distracted by what is on the TV, or else fid-dling with our mobile devices. In fact, according to the statistics, almost half of us watch TV while eating, and many of us do so in separate rooms from our companions! One ingenious solution to counteract all this technology-induced distraction was trialled by Brazilian bar Salve Jorge back in 2013. They introduced the 'Offline Glass', a beer glass whose base is partly cut away so that it can only stand up if supported by the customer's mobile device. The idea was that this enforced 'disconnection' from the patron's technology would hopefully result in people being more sociable when out drinking.

I am sure that we have all seen those unromantic couples dining together, not speaking to one another, engrossed instead in whatever happens to be on their mobile screens. Dining together alone, as it were. Of course, sometimes even those who aren't distracted by their hand-held technology don't necessarily have anything much to talk about. One idea to alleviate this situation comes from the Bocuse res-taurant at the Culinary Institute of America campus in upstate New York, where they provide a box filled with cards on every table. Each card poses a culinary question or joke. But why are they there? It is certainly not something that you would normally expect to see in a fancy restaurant, is it? When I asked, the last time I visited, the insti-tute's director told me that these games had been strategically introduced to help break the ice for those couples who may be lost for anything to say. The hope is that the cards will improve the diners' mood, and so enhance their enjoyment of whatever they happen to order. This is another example of a mental palate cleanser, like the plastic cow we came across earlier.

Eating with the TV on is one of the worst things you can do in terms of increased consumption. Finding that people eat 15% more food with the telly on as compared to when it is off is not unusual. That said, not all shows are equally bad for our waistlines. It seems to depend on how engaging the TV show is, and whether or not we have seen it before. For instance, Dick Stevenson and his colleagues in Australia found that women who viewed the same episode of *Friends* twice consumed significantly more food than those who got to watch different episodes of the hit TV show instead. Generally speaking, the more food-related sensory cues we are aware of, the less we tend to eat. Hence, when distracted by the television (and presumably this also applies to mobile devices), the danger is that we simply fail to pay attention to the food-related stimulation and so end up consum-ing more before we realize that we are actually full. And there are other reasons not to watch TV while eating, as this no-nonsense advice suggests: 'Mealtimes have been noted as one of the most com-mon times children communicate with parents, so if possible, guard your mealtimes from outside distractions. Turn off the TV and cell phones.'

Do you enjoy dining alone?

I have a few colleagues, mainly, it has to be said, chefs, who say that they sometimes (only sometimes, mind you) prefer to eat alone. Why? Well, because it allows them to really pay attention to what is on their plate (i.e., to the flavour combinations and texture contrasts). If they are going to a gastronomic hotspot, they may well choose to dine solo rather than risk being distracted by the need to make conversation. Perhaps I should have used this line the time I turned up by myself for a meal at The Fat Duck. Call me unromantic, but I hadn't even registered that it was Valentine's Day!* Needless to say, Heston has enjoyed teasing me about this whenever the topic has come up since.

For me, like many others, I suspect, the experience of dining alone is never anything like as enjoyable as dining with others, no matter how enjoyable the food. After all, great food and drink are nearly always pleasures best shared, and the better the meal or wine, the more you want to share it. To the extent that our mood is likely to be better when we dine in company than when sitting by ourselves, I am sure that the food and drink really does taste better to us too when in company (at least when in company we enjoy). Intriguingly, dramatic mood swings are associated with significant changes in taste and smell sensitivity. Our hedonic response to food and drink can also be affected. Who, after all, ever had a great-tasting meal when fighting with their partner?

The shared meal is a universal human phenomenon, with evidence of feasting going back more than 12,000 years in the archaeological record. And there is little more expressive of companionship – derived from the Latin *cum*, 'together', plus *panis* 'bread' – than the shared meal. Carolyn Steel, in her book *Hungry City*, makes the point that '[w]e are hard-wired to feel close to those with whom we share food, and to define as alien those who eat differently from us'.[4] (She also picks up on a great line from Oscar Wilde's *A Woman of No Importance*: 'After a good dinner one can forgive anybody, even one's own relatives.') The latest research shows that eating together can increase agreeableness too – thus giving a whole new slant to the topic of gastrodiplomacy.

According to my Oxford psychology colleague Professor Robin

---

* In the good old days, the solo diner would distract themselves with a book. Nowadays, of course, it is increasingly the smartphone that offers companionship at the table. However, being one of the only people left on the planet without a mobile phone, that option unfortunately wasn't open to me.

Dunbar: 'The act of eating together triggers the endorphin system in the brain and endorphins play an important role in social bonding in humans. Taking the time to sit down together over a meal helps create social networks that in turn have profound effects on our physical and mental health, our happiness and wellbeing and even our sense of purpose in life.'[5] All the more worrying given the latest statistics showing that almost 70% of people have never shared a meal with their neighbours. Furthermore, when quizzed, 20% of us admitted that it had been more than six months since we last shared a meal with our parents.[6] It should never be forgotten that: 'The table is the original social network.'[7]

Gastrophysics research shows that dining with companions can, and probably does, exert a significant impact on how much we eat. Whether we eat *more* or *less* depends on who we are with, and how much we are trying to impress them. Evidence from both the laboratory and from more naturalistic dining studies shows that the amount of food consumed typically goes up when we dine with others, as compared to when we eat alone. This increase is more pronounced when dining with friends and family than when we are with those with whom we are less familiar. Males, in particular, tend to eat more in groups than when alone in the restaurant setting. These social effects on consumption may, in part, relate to the longer duration of a dinner in company. However, if we are trying to impress whoever we are with (or else are nervous!), then we may well end up eating less. We also tend to consume less than we otherwise might when those around us hardly touch their food. Remarkably, such social effects on consumption have even been demonstrated in subjects who have had nothing to eat for twenty-four hours.

The next time you go out for a meal you should bear the following in mind. The biggest problem with dining in company is the fact that it reduces the likelihood of you getting to order first. This matters because those who order first generally tend to enjoy their food and drink more than those who order later. The latter often end up feeling that they should have chosen something different and, as a result, they may enjoy the experience a little less than they would have done had they chosen first. And given that women tend to order first in mixed company, this probably means that they generally enjoy their meals out a little more than men.

As we saw in the 'Sound' chapter, many people have been complaining in recent years about restaurants and bars that are so noisy that they simply can't hear themselves think, never mind taste the food. As one commentator put it: 'You go to restaurants to be social. These days, you often come out none the wiser of what the other person has said.'[8] As we have seen, one response to this has been to go to the other extreme and introduce silent dinners, where none of the diners are allowed to speak. However, it is the fact that the meal is fundamentally a social activity/occasion that helps, I think, to explain the long-term failure of this particular concept. Putting in earplugs to drown out the noise, or donning a pair of headphones to play a dish-specific soundscape or piece of music, works well for one course, but any more than that and the social dynamics of the meal will be disrupted too much.

Notice, in contrast, that while all those dine-in-the-dark restaurants also remove one of the diner's senses, the darkness doesn't mess with the social aspects of dining. This is the key difference between dark and silent dining. If anything, diners are likely to have *more* to talk about when the lights go out – for instance, comparing their uncertainty about what exactly it is that they are eating.

Catering for the solo diner

Until recently, if you saw someone dining alone, they might have seemed a little sad, almost a social outcast. 'What, don't they have any friends?' you might have thought to yourself. This stigma, though, is starting to wane. Indeed, more people are dining out by themselves than ever before, with the number of solo diners more than doubling in the two years to 2015. In fact, a party of one is the fastest-growing size for reservations at restaurants in the UK. But what do all these solo diners do to keep themselves entertained while waiting for their order to arrive? According to one recent survey, 46% of people said that they would pass the time reading a book, while 36% played with their phone.[9]

The changing sentiment is nicely captured by the following quote from someone writing to the BBC after the publication of an article on solo dining: 'I remember a time, only a few years ago, when I found the idea of eating out alone to be a depressing prospect. I would view solo diners as sad and lonely people. Now, I eat out by myself quite often, and sometimes prefer it to the company of others. One thing more than any other has made this change from my perspective – a smartphone. I suppose solo diners really aren't solo any more at all.'[10] Some even view those individuals who dine solo in public as being confident high-achievers, out enjoying the well-deserved rewards of their labour, like the wonderfully opinionated food critic Jay Rayner: 'I'm not worried about anyone thinking I'm a sad bastard [. . .] Eating alone should be dinner with someone you love.'[11]

Part of this change in attitude may relate to the fact that more diners are now sharing every detail of what they eat, and pictures of their meals, on blogs or social media. There is even a collection of images on Tumblr called 'Dimly Lit Meals for One'. This recent trend seems to be gaining traction and goes hand in hand with our growing connection with our mobile devices, a relationship that, according to some, is best described as nothing short of a 'love affair'. Others, meanwhile, are happy to isolate themselves with their MP3 players and over-ear noise-cancelling headphones.

Many forward-thinking restaurateurs see this shifting demographic as a marketing opportunity, like Ivan Flowers, the executive chef at San Diego's Top of the Market restaurant. He was brought in by the management specifically to increase the number of solo diners: '[They] felt that while the eatery already had bar seating in front of the open kitchen, it was underutilized because the chefs weren't interacting enough with the customers.' Flowers continues: '[S]olo diners sitting by the kitchen now get "to see a show", which includes cooking demos, free tastings and conversation with the chefs.' Some newspapers have now even started to make recommendations for solo diners in their restaurant reviews.[12]

Eenmaal, a pop-up restaurant in Amsterdam only features tables for one. Who would have guessed that such a venue would have been booked solid for a year since opening its doors? No surprise, then, that those behind this venture are currently planning to expand, with branches planned for London, Berlin, New York, Antwerp and beyond. In the words of Marina van Goor, the project's designer: 'I noticed that in our society, there is no room for being alone in a public space, unless you are going somewhere.'[13] My suspicion is that changing the layout of the restaurant so that it caters solely for individual diners will never be anything more than a niche offering, given how much we still rate the eating experience more highly when in company. That said, many restaurateurs could nevertheless do much more to adapt what they offer to this change in our eating habits.

Tapas-ization

Even as solo dining is on the rise, statistical analysis of the language of restaurant menus reveals a marked increase in the use of sharing terms. Nowadays, one is far more likely to see charcuterie boards, tapas and mezze plates appearing on the menu, all, note, dishes designed for sharing. They are also more informal – another popular trend in dining currently. Here in Oxford, for example, at our top gastro-pub, the Magdalen Arms, many of the dishes on the menu are for two, three, four or even five diners to share.

And then there is the rise of the tasting menu – lots of smaller plates chosen by the chef. More often than not, everyone at the table has to agree to order these. As a gastrophysicist, I would advise the restaurateur to seat diners at a round table if that is what they want customers to do, because people are more likely to feel like they belong when seated at a round table than at a square one. By contrast, those at an angular table tend to show more selfish traits in group settings. All of a sudden, the fact that Chinese banquets are always held around a circular table starts to make much more sense. And going even further back, don't forget about King Arthur and his Knights of the Round Table. The main reason for seating diners at angular tables is to maximize the number of diners in a space. That said, while the round table is certainly the most democratic solution, if it is too large, it can make conversing with those on the other side of the table awkward.

It is not just the plates that restaurateurs want you to share these days. You have probably come across the communal/informal dining concept whereby everyone gets to sit at large long tables. This approach constitutes a distinctive aspect of the design of restaurants like Wagamama and Busaba Eathai (both incredibly successful ventures from top restaurateur Alan Yau). The Pain Quotidiens chain does much the same. In some sense, anyone who chooses to frequent one of these informal eating establishments is sharing the table with strangers – though, as the saying goes, 'There are no strangers, just friends that we have not yet met.' The physical distance from those sitting next to you might well be the same as in other popular restaurants where they cram the tables for two in against the banquette. And yet I believe that there is still something qualitatively different about 'being connected', parked at the same long table. In fact, I think that I can feel a new gastrophysics experiment coming on to get to the bottom of the matter! Just how much do we enjoy the experience when crammed at a long table with strangers sitting right next to us? As always, there is definitely more research to be done here.

Have you ever thought about how strange dining out is?

Think about it carefully, and you begin to see how strange all this eating in public in close proximity to strangers really is. How, for example, might someone from another culture view the social aspects of dining in a restaurant if they were to visit our Western twenty-first-century world? Perhaps much like Antoine Rosny, a Peruvian traveller (despite his French-sounding name), who described his experience on first visiting a Parisian restaurant back at the start of the nineteenth century (i.e., in the very early days of the restaurant): 'On arriving in the dining room, I remarked with astonishment numerous tables placed one beside another, which made me think that we were waiting for a large group, or perhaps were going to dine at a table d'hôte ('host's table'). But my surprise was at its greatest when I saw people enter without greeting each other and without seeming to know each other, seat themselves without looking at each other, and eat separately without speaking to each other, or even offering to share their food.'[14]

One intriguing performance art piece that probes the meaning of sharing a meal comes from Indonesian artist Mella Jaarsma (see Figure 7.2). Members of the public (anywhere between two and six of them) are invited to put on a bib from which a flat table surface is suspended. Diners pair up and then both order for and subsequently feed each other. This kind of intimate performance gives rise to a literally shared meal. Notice also how this wearable table temporarily binds the diners (performers), as they create a mutually supported surface from which to eat. (Though I do worry about what exactly happens when one of the diners needs to visit the toilet!) The comment of one participant who took part is interesting: 'In enacting Mella Jaarsma's piece, I experienced feeding another person and being fed by another person for the first time in my adult life. [. . . Throughout the meal] one constant remained – the way the ritual of feeding and being fed articulates power relations [. . . O]ur proximity to

Figure 7.2. Still from a performance of *I Eat You Eat Me* (2001–12).[15]

those we have power towards makes us generous; one wishes this was the case more often outside of art.'

Artist Marije Vogelzang (from the Netherlands) has created a piece called *Sharing Dinner*, in which the diners are connected by cloth. In this visually striking installation, diners insert their heads and arms through slits in a white tablecloth suspended from the ceiling (see Figure 7.3). As the artist notes: 'I used a table with a tablecloth, but instead of putting the cloth on the table, I made slits in it and suspended it in the air, so that the participants sat with their heads inside the space and their bodies outside. This physically connects each person: If I pull on the cloth here, you can feel it there. Covering everyone's clothing also created a sense of equality. Initially I was concerned that people would reject the experience, particularly because the participants didn't know each other beforehand, but it actually increased their desire to relate to one another, and brought

Figure 7.3. *Sharing Dinner* by culinary artist Marije Vogelzang (Tokyo, 2008).

about a feeling of being in something together.'[16] Vogelzang also uses food to encourage sharing. She would serve one person a slice of melon on a plate that was cut in two. Meanwhile, the person sitting opposite would be given ham on a similar plate. This resulted in the participants (many of whom didn't know each other) naturally sharing their food to make the classic combination.

The telematic dinner party

Many of us who are fortunate enough to have a family to dine with nevertheless still sometimes find ourselves away from home, when travelling for work, say. As a result, we can all too easily miss out on the shared time typically centred around the dinner table. Indeed, this has been identified as a growing problem by researchers working in the field of human–computer interaction, a number of whom have started to investigate whether technology can somehow be used to reconnect people who find themselves far apart, allowing them to share meaningful (i.e., embodied) virtual mealtime experiences. Thus, the idea of the 'telematic dinner party' is born – just think of it as 'Skeating', i.e., Skyping while eating.

While the concept is undoubtedly an intriguing one, there are a number of challenges for anyone wanting to make a successful technological intervention in this space. For example, what happens when those who are dining together digitally eat different foods – does that hinder their making 'a connection'? Another important issue here concerns how to make this kind of virtually shared dining experience more immersive and engaging. As a guest at a preliminary telematic dinner party said: 'I don't feel that I shared food with them. It felt like we [were] together in one room and they were eating in another room. There was no sense that we were sharing.'[17] This is definitely not what tomorrow's experience designers want to hear. Dining together involves the tight (albeit unconscious) synchronization of the diners' behaviour, so that trying to replicate the precisely coordinated choreography at a distance, where a lag can sometimes delay the signal coming from the other end, is likely to disrupt the more communal aspects of the meal.

I could, just possibly, imagine this kind of technological solution having value in extreme circumstances, for someone on a long space mission to Mars, say, who wants to reconnect to their family back on planet Earth. However, stuck down here on the ground, I really can't work up any enthusiasm for the telematic dinner party concept, even assuming it worked perfectly (i.e., with no synchronization issues). I'm tempted instead to place my bets on a different technological innovation. I am thinking of the various meal-sharing apps that have been launched over the last couple of years. For a small fee, they allow those who find themselves alone in an unfamiliar place to dine with a local in their own home; so a meal is physically shared, just not with the subscriber's family. The various sites offering this kind of service each have a somewhat different feel, thus there is most likely something for every taste – and if there isn't yet, you can guarantee there will be soon! For instance, the US site EatWith has the feel of a supper club, while UK-based VizEat emphasizes the opportunity to eat with locals, to get inside a culture by eating it, as it were. According to one of VizEat's co-founders, Camille Rumani, this site had attracted more than 170,000 hosts in 115 countries within a couple of years ago of starting up (in July 2014).

The question that we should probably all be asking ourselves is whether such meal-sharing apps will come to revolutionize the way in which we dine when we are away from home in much the same way that Airbnb has transformed where we stay and Uber has transformed transportation. (Note here that an UberEats app was recently launched in cities across the US, with the promise to make 'getting great food from hundreds of restaurants as easy as requesting a ride'.) According to Euromonitor International, a market intelligence agency, 2015 was *the* year to look out for 'peer-to-peer dining' as one of the biggest trends. This involves direct interaction between cook and diner, without the mediation of a restaurant (chain) – just think of all those chefs who are starting to cook from home as symptomatic of this trend. But, I would argue, by far the biggest potential market is those individuals who haven't left their own home, and yet still have no one to dine with. One recent start-up connecting them is Tablecrowd, which combines eating with social networking; another is Tabl, which offers essentially curated social dinners in the south of England. Dining together is such a primal urge, so next time you get peckish, why not invite someone to eat with you? Most likely, you will end up having a better time than you would on your own. Just remember to order first, though, if you want to maximize your enjoyment!

8.   Airline Food

It was back in 2014 that I first got to thinking about the food and drink while in the sky. Sitting on yet another long-haul flight when my laptop battery finally died, I ended up watching the cabin attendants slowly wheeling the drinks trolley down towards the back of the plane. And it was then that it struck me – just how many people order a tomato-juice-based beverage while up in the air. Now, though you might see the odd person ordering a Bloody Mary down on the ground, it really is a pretty rare occurrence, at least in the circles in which I move. However, up above the clouds, watching the drinks flying off the trolley, it really felt like every fourth order or so involved tomato juice. But what exactly is so special about the red fruit (or is it a vegetable?) in the air, and how might the insights garnered from the drinks trolley in the sky lead to a radical redesign of airline food?

Well, firstly, you shouldn't just take my word for it. Before going any further, we need to check on the veracity of the observation. Luckily, my intuitions proved correct: it turns out that tomato juice makes up 27% of the drinks ordered in the sky. What is more, there is a whole section of the population who regularly order a tomato juice from an air steward or stewardess but who would never think of doing so with their feet planted firmly on the ground. Of the more than 1,000 passengers questioned in one survey, 23% fell into the latter category. So what exactly is going on here? Before answering that question, though, let's take a quick look at the history of airline food.

The way it was

Plane food hasn't always been bad. Back in the early days of commercial flight, the airlines would put on quite a spread for anyone who was rich enough to fly. Believe it or not, they used to compete on the

Figure 8.1. Airline food as it used to be! 'How many lobsters would you like today madam?' Passengers being served fresh Norwegian lobsters (with their shell still on). Note the aperitifs set on ice too.

quality of their food offering, with a carvery, lobster, prime rib, etc. available to all who wanted it (see Figure 8.1). Perhaps this helps explain why the pop-up dining concept Flight BA2012, in Shoreditch, East London, was such a success. The hipsters of Hoxton were able to sample a three-course meal inspired by the airline's 1948 first-class menu. It is hard to imagine that contemporary airline food would have had quite the same appeal (see Figure 8.2).

Everything changed in 1952, though, with the introduction of economy class and the associated economies of scale required once passenger numbers increased dramatically. The International Airline Transport Association (IATA) didn't help matters, either: they actually brought in guidelines limiting what food could be offered up in the skies, at least in economy class. Scandinavian Airlines were even fined

- **MENU** -

COPENHAGEN–BANGKOK

SALADE NICOISE
Mixed Salad "Niçoise"

CREPES FORESTIERE AU BACON
French Pancakes Filled with Mushrooms and Bacon

CREME BAVAROISE AU RHUM
Rum Bavarian Cream

CAFE
Coffee

AVANT L'ATTERRISSAGE/Before landing

FRUITS DE SAISON
Assorted Fresh Fruit

PETITS PAINS ET CROISSANT
Rolls and Croissant

BEURRE, FROMAGE ET CONFITURE
Butter, Cheese and Marmelade

OEUFS POCHES A LA FLORENTINE
Poached Eggs on Sautéed Spinach with Mornay Sauce

POMMES DE TERRE ROESTI
Roesti Potatoes

SAUCISSES
Sausages

CAFE OU THE
Coffee or Tea

BANGKOK–SINGAPORE

HORS-D'OEUVRE VARIES
Assorted Appetizers

POISSON A LA THAILANDAISE
Poached Fish Thai Style in Black Bean Sauce

LEGUMES CHIANG MAI
Mixed Chiang Mai Vegetables

RIZ FRIT
Fried Rice

OU/or

COTELETTES D'AGNEAU A L'ITALIENNE
Lamb Chops with Tomato and Oregano Sauce

POMMES DE TERRE A LA BOULANGERE
Potatoes à la Boulangère

EPINARDS AU BEURRE
Battered Spinach

FROMAGE
Cheese

TARTELETTE AU CITRON
Lemon Meringue Tartlet

CAFE
Coffee

PLEASE ACCEPT OUR APOLOGY IF, DUE TO PREVIOUS
PASSENGER SELECTIONS, YOUR CHOICE IS NOT AVAILABLE.

IN ADDITION TO OUR FRESHLY BREWED COFFEE,
WE SERVE TEA AND DECAFFEINATED COFFEE WITH ALL MEALS.

Figure 8.2: An impressive-looking French menu served on board a flight from Copenhagen to Singapore, stopping in Bangkok on the way, in the 1950s.

$20,000 for serving their transatlantic passengers a bread roll that was deemed to be just too good, following a complaint from Pan Am, one of their competitors. In recent times, though, there can be no doubt that (in real terms) less and less money is being spent on the food offering in the air, assuming, that is, that any food is provided at all.

Once upon a time, the food was pretty much the only thing that kept air passengers from thoughts of their own demise, should the plane's none-too-reliable engines fail. Hence the quality of the offering was especially important, given that the passenger had little else to occupy their minds other than admiring the view from the window. Nowadays, though, everything has changed. For one thing, plane travel is, thankfully, safer than ever. What is more, all manner of entertainment is available to passengers at the touch of a button.

Mind you, the cabin atmosphere up at 35,000 feet isn't all that conducive to fine dining. The lowered air pressure, together with the lack of humidity (the air, don't forget, is recycled through the cabin every 2–3 minutes), really doesn't help, with food and drink losing roughly 30% of its taste/flavour when sampled at altitude. Aware of these problems, a number of the airlines test their menus under conditions that mimic the atmosphere up there down here, if you see what I mean. In Germany, for instance, at the Fraunhofer Institute, half an old Airbus plane has been plonked in a low-pressure chamber and where they test people's reactions to the foods that they are thinking of serving up above the clouds.

More often than not, though, the airlines have opted to load the food they serve with ever more sugar and salt, to enhance the flavour. No surprise, therefore, that the airline food served these days isn't the healthiest. In fact, it has been estimated that the British consume more than 3,400 calories between their check-in at the airport and their arrival at their destination.

Over the years, the airlines have sought advice from chefs to help improve their food offering. Back in the day, it was French chef Raymond Oliver who was brought in by Union de Transports Aériens (the precursor of what was to become Air France). His advice fundamentally changed what was served in the air, and soon established the standard format for the meals that so many of us have now become all too used to. In fact, the chicken or fish dishes that one finds on many of today's economy-class menus can be traced directly back to the chef's early suggestions. For instance, Oliver recommended serving foods that the passengers would find familiar; not quite comfort food, exactly, but at least something reassuring. He was looking for hearty meals that were easy to prepare and heavy to digest, the idea being that then the passengers wouldn't get hungry again before the plane landed. The meals shouldn't lose too much of their flavour when reheated either. The chef's suggestions: coq au vin, beef bourguignon and veal in cream sauce. (It was 1973, after all!) These dishes also had the advantage that the meat, drenched as it was in sauce, wouldn't dry out too much when heated up in the plane.

Can celebrity chefs really cut the mustard at 30,000 feet?

Today, it's much more common for airlines to bring in a chef to try to improve their food offering. And a number of them have hired celebrity chefs in order to spruce up their airline meals, Neil Perry from Australia teaming up with Qantas, for instance, or Heston Blumenthal with British Airways, or the late, great Charlie Trotter advising United Airlines. (Trotter's top tip was short ribs spiced with Thai-style barbecue sauce; the addition of spice and the inclusion of sauce are both good ideas when it comes to dining at altitude.) Meanwhile, Air France currently has so many great chefs to choose from that they rotate their affiliation on a regular basis.

My guess is that even those sitting in the front of the plane would have little inkling of the star chefs' involvement in the dishes that they are tucking into were it not for their names emblazoned on the menu cards. Certainly, I have yet to see any evidence to support the claim that the chefs' interventions in any of the cases just mentioned actually led to a significant increase in passenger satisfaction. And, perhaps more tellingly, the airlines that have sought advice from the top chefs don't seem to appear any more frequently in the list of the top-ten airlines for food that is published annually. Regardless of how many Michelin stars a chef has, their food will never taste (quite) as good in the air as when served in their flagship restaurant down on the ground. What many of the frequent business passengers do appear to appreciate, though, is the change in format from the fixed-course meal service of old to more of a grazing approach, allowing them to eat what they want, more or less when they want – food on demand, as it were.

As we have already seen, to a large extent what we think about food and drink depends on the context or environment in which it happens to be consumed. And airline food is no different. One of the other significant impediments to progress in the sky relates to the long-term catering contracts that many of the airlines have signed up to. Thus, even if the airline, or the chef that the airline has brought in, wants to change the culinary offering, it can prove difficult to do so in practice. Behind the scenes, a number of the catering suppliers have now started to bring in some innovative chefs to help advise them directly. The more fundamental issue here, though, is that the chefs' input is typically restricted just to the ingredients, recipes and preparation of the food. And as we are about to see, any solution that is solely focused on the food will only take you so far. It's time to bring in the gastrophysics perspective.

What's the link between the humble tomato and aircraft noise?

Let's get back to the juice! Once the plane has reached cruising altitude, the passenger's ears will be exposed to somewhere in the region of 80–85 dB of background noise, depending on how close they are to the engines. This racket suppresses our ability to taste. However, it does not affect our perception of all foods equally. The really special thing about tomato juice and Worcestershire sauce (both ingredients in a good Bloody Mary) is umami, the proteinaceous taste, experienced in its purest form as monosodium glutamate or MSG. While it has long been popular in East Asian cuisine (e.g., in Japan, where the term translates as 'delicious', 'savoury' or 'yummy'), it has recently started to attract the interest of chefs from many other parts of the world too. In the West, those foods that you are likely to have come across that are rich in umami include Parmesan cheese, mushrooms, anchovies and, of course, tomatoes. So, could this help to explain the mystery of why so many people order a tomato-juice-based beverage while up in the air?

Well, in 2015, researchers from Cornell University finally got around to assessing the effect of loud aeroplane noise on people's ability to taste umami. Participants sitting in the lab had to rate the strength of a series of clear drinks, each containing one of the five basic tastes presented at one of three different stimulus concentrations. Each of the solutions was tasted in silence and also while listening to pre-recorded aeroplane noise at a realistic decibel level. Intriguingly, the perceived intensity of the umami solutions was rated as significantly higher when the background noise was ramped up. By contrast, ratings of sweetness were suppressed, while ratings of the taste of the salty, sour and bitter solutions were unaffected. Given such results, British Airways' decision to introduce an umami-inspired menu on their flights back in 2013 starts to make a lot more sense.

But why should loud noise affect some tastes but not others? Well, one intriguing theory is that our responsiveness to different tastes varies as a function of how stressed we are – this being just what many passengers may be feeling while flying, especially on a bumpy flight. In one older study, for example, sweet, but not salty, solutions were rated as significantly more pleasant under conditions where stress had been induced by the presentation of loud noise. One suggestion that has been put forward to try to explain this surprising result is that the energy that is signalled by sweetness might be just what an organism needs when it comes to dealing with the kinds of situation that give rise to stress in the first place. Presumably, a similar evolutionary story could be told about the increase in the perceived intensity of umami under conditions of loud noise. For, just like sweetness, umami is also a nutritive tastant, signalling, as it does, the likely presence of protein. But whatever the correct explanation ultimately turns out to be, the key point remains that loud noise generally suppresses sweetness and sometimes saltiness, while at the same time enhancing the taste of umami.

What happens when people are given proper food to taste, rather than just pure tastants dissolved in solution? Well, playing loud white noise – think of the static sound on an untuned radio – leads to ratings of both the sweetness and saltiness of various snack foods such as potato chips, biscuits and cheese being suppressed. Somewhat surprisingly, though, crunchiness ratings were actually higher when the background noise was turned up (as compared to silence). Perhaps, then, the airlines should be thinking about adding more crunch to the food that they serve, and adding more of the other noisy textural food attributes, such as crackly and crispy too. This is likely to have the added advantage of improving the perceived freshness and palatability of the food. Indeed, this is why having a bowl of fresh fruit available (as some airlines do for business-class passengers) is a good idea. And sprinkling the salad with sesame seeds to boost the crunch factor would be a darn sight cheaper than hiring one of the world's top chefs.

So, counterintuitive though it may sound, donning a pair of noise-cancelling headphones could actually be one of the simplest ways in which to make the food and drink taste better at altitude. But, now that we have got rid of the background noise, the next question is: what else, if anything, can you listen to, in order to make your food taste better?

Supersonic seasoning

Late in 2014, British Airways introduced 'Sound Bites' for their long-haul passengers. Once they had chosen their in-flight meal, passengers could tune in to one of the channels in the seatback entertainment system. There they would find a carefully chosen playlist of popular tunes that had been specially selected to complement the taste of the food. The musical selections were based, in part, on research findings from my lab. A number of the tracks were chosen to boost the perceived authenticity/ethnicity of the dishes, given research showing that this attribute can be enhanced by presenting matching music (or, for that matter, any other sensory cues) appropriate to the region associated by people with the food (see 'The Atmospheric Meal'). Think lasagne, or pasta, while listening to one of Verdi's arias (and if you can find a red and white checked tablecloth, even better), or perhaps Scottish salmon with The Proclaimers?

Some of the first empirical evidence supporting the existence of sonic seasoning came from research conducted with The Fat Duck Research Kitchen in Bray. Together with the then head research chefs, Steffan Kosser and Jockie Petrie, we were able to demonstrate that listening to soundscapes containing lots of tinkling, high-pitched notes accentuated the sweetness of a bittersweet cinder toffee, whereas listening to low-pitched noises brought out the bitterness instead. The effects, it should be said, weren't huge (5–10%), but they were large enough to potentially make a difference to the tasting experience while up in the air. So why don't you forget about adding sugar the next time you are eating at altitude, and instead just tune in to some sweet, calorie-free music? That said, while we have now got some pretty effective sweet tunes, we are still struggling to create the perfect sonic salty backdrop.

Now, I am assuming that you have taken my advice and got yourself a pair of noise-cancelling headphones, and that you are listening to the right sort of music to complement, and thus improve, the flavour of whatever you happen to be eating or drinking. So, what next? What else can be done to improve the meal experience while passengers are up in the air? Well, if the food really is worth savouring, one simple tip here would be to pause the movie. For according to the ground-based research covered in the 'Social Dining' chapter, you ought to find that you enjoy your food a little more while, at the same time, finding yourself satisfied with less of it.

Tasting under pressure

In addition to the background noise, one of the other problems with tasting at altitude is the reduced cabin air pressure. These days, planes are pressurized to create an atmosphere that is equivalent to what one would find at an altitude of approximately 6,000–8,000 feet. Under such conditions, it becomes harder to taste sweet, sour, salty and bitter. No wonder then that airline food tastes so bad. However, the more profound problem is that the number of volatile aromatic molecules in the air also decreases as the cabin air pressure drops. It is this that can really suppress our flavour perception. One innovative solution might be to wear a Breathe Right nasal strip. These plasters were originally designed for athletes to place over their nostrils in order to increase the intake of air, potentially enhancing their sporting performance. Wearing one can lead to nasal airflow going up by as much as 25%. So thinking laterally here, one might consider providing the passengers with one along with their earplugs on boarding the plane, as a means of increasing their exposure to any of the volatile aromas of food and drink that are still floating around in the atmosphere at altitude. However, while this solution has yet to be tried in the skies, the findings from research conducted down on the ground have unfortunately so far been, how shall I put it, less than encouraging.

Another recommendation for improving the tasting experience while up in the air (i.e., at low atmospheric pressure) comes from Professor Barry Smith, a philosopher and wine writer working out of the University of London. He has noticed that high-altitude wines (e.g., New World Malbec from Argentina) tend to be rated better in the air than one might have anticipated from tasting them on the ground. Why should that be so? Well, his suggestion is that the atmospheric conditions on the side of the mountain where many of these wines are made (i.e., blended) is, in some sense, closer to those one finds in an aeroplane cabin than for some other wines. The grapes that go into Argentinian Nicolas Catena's Zapata wines, for instance, are grown at around 5,700 feet. So perhaps it is no wonder his wines taste better at altitude. So next time you get the chance to choose your wine on a plane, remember another one of Smith's suggestions and go for a fruitier number. Whatever you do, Smith argues that you'd do well to avoid those prestige wines with firm tannins, as they may well leave you with a fiercely astringent/bitter taste in your mouth.

Another problem with the atmosphere in a plane cabin is that the humidity levels are much lower than on the ground (below 20%, as compared to 30% or more in the average home). The good news again, at least for those travelling in style, is that the humidity apparently tends to be a little higher at the front of the plane. Lower humidity levels also impair our ability to taste, since they tend to result in the drying of our nose, making it harder to detect the remaining volatile odour molecules. A few years ago, chef Heston Blumenthal came up with his own idiosyncratic solution to this particular problem. His recommendation was that anyone wanting to enjoy their food and drink more while flying should give themselves a nasal douche with a water spray. The idea here (perhaps a little tongue-in-cheek) was to try and increase the humidity in the nostrils to make up for the lack of humidity in the air that is recirculated every couple of minutes or so through the cabin. With all due respect, though, while the suggestion undoubtedly makes for engaging television, it is hard to imagine anyone actually taking the advice seriously. And anyway, before getting too carried away, just remember that should the nasal douche work as hoped, it would also increase your ability to smell the passengers sitting close by. Are you sure that is really what you want?

Simple tips for service

For my brother, a self-confessed wine buff, the realization came while staying in a Swiss ski chalet a few years ago. He had decided to open a long-treasured bottle of wine, but the only thing that he could find to pour the wine into late one night was the plastic water cup from the bathroom. He knew exactly what the wine – a Kistler Chardonnay – *ought* to taste like. He had, after all, liked it so much that he had bought a couple of cases of the stuff. Somehow, though, with the wrong drinking vessel, my brother just wasn't able to recreate the great tasting experience that he knew he should be having – one that, moreover, he desperately wanted to be enjoying, given how much he had paid for the wine in the first place. On a plane, served an unfamiliar vintage, it is going to be even harder for any of us to discount the glassware and focus instead solely on the quality of the contents and how much we are enjoying the experience. Haven't we all been disappointed while drinking something expensive from a flimsy plastic container? No matter how prestigious the contents, the cheap feel of the vessel takes away from the pleasure of the experience.

What all of us know intuitively, I think, and the research data now backs up, is that we like beverages more when they are served in the appropriate receptacle than when served in something inappropriate. Just think about it, would you enjoy drinking tea out of a wine glass? Of course not! Knowing this, one really has to wonder what on earth many of the airlines are thinking: How, I ask you, can anyone honestly justify serving the complimentary glass of champagne that many business-class service encounters start with from such a light and flimsy cheap plastic glass. While serving the champagne from a plastic flute would help a little, I would suggest that anyone hoping to optimize the tasting experience should really be using glass, not plastic – for weight is crucial to the experience, no matter whether we find ourselves in the air or on the ground.

Back in the days of supersonic flight, every gram counted (much more so than it does on regular flights). At some point, designers were brought in to help develop some stylish new extra-lightweight cutlery for Concorde (plastic, while undoubtedly very light, was obviously not an option) and created some beautiful titanium cutlery: exquisite to look at, and lighter than anything metal that had gone before. Job well done! Or so one imagines they must have thought. The problem, though, was that people simply didn't like it. When trialled, it just felt too light and consequently it was never introduced on board.

Finally, I know of one innovative airline that has recently been thinking about the material properties of the cutlery they hand out. Why would they want to do that? Well, because the material from which the fork and especially the spoon (i.e., those items that enter your mouth) are made can modify the taste of the food. We conducted some research relevant to this a few years ago, together with the Institute of Making in London. We were able to demonstrate that an everyday sample of yoghurt, to which a small amount of salt had been added, was rated as tasting saltier when eaten with a stainless steel spoon that had been electroplated with copper or zinc. Such results raise the question of whether novel cutlery designs like this could be used to help season the food served in the air. Remember here that it is primarily sweet and salty tastes that are suppressed by all that loud background aeroplane noise. Unfortunately, though, while there are certain metals that can be used to bring out the salty, bitter and sour taste in food, I am not aware of any metal that can boost sweetness. Well, there is lead, I suppose; it is just that it is also poisonous. So, not quite what one wants in one's cutlery.

Will multisensory experience design really take off?

All well and good, I hear you say, but will anything actually change in the near future? What will the airline meal of tomorrow look like? The good news, according to my sources, is that one of the big airlines plans to launch an in-flight food and beverage offering that will put everything that we have become accustomed to in recent years to shame. I'm afraid I can't say any more just yet. But where one airline leads, others may, sooner or later, follow, and if this is the case, then I am hopeful that we might finally see a return to the early days of flight, when the fledgling airlines competed on the quality of their food service offering (for their admittedly deep-pocketed passengers).

Does this sound unbelievable? Well, before you dismiss it, do allow me to paint a picture of air travel as it was back at the end of the 1960s. At that time, Trans-World Airlines started running themed 'foreign accent' flights between major US cities. Let me quote Alvin Toffler, former editor of *Fortune* magazine and author of the bestseller *Future Shock* directly (for otherwise, I fear, you will think I am making it up): 'The TWA passenger may now choose a jet on which the food, the music, the magazines, the movies, and the stewardess's outfits are all French. He may choose a "Roman" flight on which the girls wear togas. He may opt for a "Manhattan Penthouse" flight' – the mind boggles – 'or he may select the "Olde English" flight on which the girls are called "serving wenches" and the décor supposedly suggests that of an English pub.'

Toffler continues: 'It is clear that TWA is no longer selling transportation, as such, but a carefully designed psychological package as well. We can expect the airlines before long to make use of lights and multi-media projections to create total, but temporary, environments providing the passenger with something approaching a theatrical experience.' And before their demise TWA weren't the only ones. For a short period in the early 1970s you might even have come across a piano lounge with a fully-functioning Wurlitzer electric piano at the back of some American Airlines 747 planes. And the British Overseas

Airways Corporation (the precursor to British Airways) apparently had it in mind to provide their unmarried male passengers with a 'scientifically chosen' blind date when they touched down in London. Little surprise that the latter scheme, called 'The Beautiful Singles of London', was scrapped when the government-owned airline came under criticism from Parliament.[1] Ultimately, then, the sky really is the limit for anyone willing to recognize the power of multisensory experience design.

9. The Meal Remembered

Humour me for a moment – what would you say was your perfect meal? What exactly can you remember of the experience? What you ate? Who you were with? Perhaps even more interesting: what do you know that you have forgotten? If that is all a bit too much, why not take a more recent dinner instead, the last time you went out to a restaurant, say, and answer the same questions. My guess is that, while you can probably remember where you were and who you were with, the details of the meal itself, the specific flavours and foods you ate, will be much hazier.* Unless, that is, you always order exactly the same dishes in your favourite restaurants, as I do.

No matter how good or bad a meal, it will never last more than a few hours. The mediocre ones we forget, the marvellous occasions hopefully stay burnished into our memories, bringing pleasure whenever we think about them. The really bad ones – well, all too often they stay etched in our minds as well, much though we might wish to forget them! For my brother, the sous vide liquorice salmon that he really didn't like is the dish that he wants to forget but has been unable to, even though he was served it more than a decade ago.

Our memory of a meal, at least of an enjoyable one, is where so much of the pleasure of the experience resides. It can last for days, weeks or even years. From the perspective of anyone trying to sell us a restaurant meal, this is important because it is a major factor in our decision whether or not to return to a particular venue or chain. Our flavour memories also play a crucial role in our decision to stick with one brand or switch to another while browsing the supermarket aisle;

* And if you are anything like me, you probably find that you have forgotten quite what your main course was by the time the dish arrives. This is why it can be so important for the waiter to explain the dish when they bring it to the table, especially if it is a non-standard offering.

once again, such a decision is often based on our recollection of the taste of the product, or what we thought about the experience the last time we encountered it.

What is a food memory?

The simplistic view here would be that our recollection of a meal is merely a weaker version of what happened at the time – 'devoid of the pungency and tang', as William James, one of the godfathers of experimental psychology (and brother of novelist Henry), once so evocatively put it. However, the gastrophysicist knows only too well that our minds play tricks on us. Not only do we forget altogether certain of the things that we experienced so vividly even just a short time ago but we also misremember and confabulate others. More often than you might imagine, we recollect things that probably didn't occur at all, or at least not in the way we remember them. Our recollections of meals, both good and bad, are no different in this regard.

Storing every detail of an experience (be it a meal or anything else for that matter) in memory would simply be too effortful. So our brains use a number of cognitive shortcuts to help. For instance, we tend to keep track of the high and low points (the peaks and troughs), and how meals start and end (termed 'primacy and recency effects'). As a result of another shortcut, we also tend to neglect the duration of events that don't change much over time. The latter (known, unsur-prisingly, as 'duration neglect') has been demonstrated to apply to meals. Such mental heuristics are efficient in that they help us to recall the gist without necessarily having to remember every detail of our lives. However, which elements of a meal will stick (i.e., the end, the peak, etc.) appears to depend on the specifics of the situation.

I would argue that knowing about such 'tricks of the mind' is abso-lutely crucial for anyone wanting to deliver more memorable food and beverage encounters. So, bearing these factors in mind, it's time to call in the 'experience engineers', the researchers who have made it their life's work to study what exactly sticks in our memory, and why.

The main aim here, in the gastrophysics context, is to get those you serve, whoever they might be, to remember more of the good stuff about your interaction with them. Do you, for instance, still remember the lime *gelée* incident from the start of the book? One chef in Washington, DC, Byron Brown, even created a theatrical dining experience in 2011* with the specific intention of enhancing his diners' memory of the event.

Common sense suggests that the best food and beverage designers – and here I am thinking of the world's top chefs, molecular mixologists and culinary artists – should be trying to create the ultimate tasting experiences for their customers. However, what those at the top of their game *really* ought to be focusing on is the creation of the most robust memories possible. Until you get that distinction clear, that our perceptions of food and drink while eating and drinking and our recollections of those consumption episodes differ, both quantitatively and qualitatively, you really can't hope to deliver the best long-term memories. The meal itself and our recollection of it are linked, obviously, but they diverge from one another in systematic ways that the gastrophysicists and experience engineers know just how to capitalize on.

One of the chefs with whom I work closely, and who has obviously been hanging around me too much, has been running his own experiment (just like a proper psychologist). What he wanted to know was just what his diners remembered of the fabulous meals he served. The chef emailed a questionnaire out to his guests a couple of weeks after they had visited his restaurant. He was in for quite a shock! For while those who chose to respond could certainly recall that they had very much enjoyed the experience, what they turned out to be very bad at was remembering precisely what it was that they had eaten. Funnily enough, the kind of thing that made the biggest impression in his diners' minds was, say, that time when the waitress sprayed some aroma or

---

* I was particularly pleased to see that it was one of my former students, Ed Cooke, a psychologist with a truly phenomenal memory, who was advising on this project. In the middle of tutorials, Ed would come out with truly amazing detail from books that he had read many years earlier.

other over their dish while they were seated at the table. In other words, it was the more theatrical, surprising and/or unusual aspects of service that stuck in people's minds, far more than the taste of the food itself. This failure to remember the food wasn't any comment on the quality of the chef's cuisine. The dishes themselves were delightful. Most were, one would have said at the time, quite memorable. However, it turned out that they weren't actually remembered that well after all, or at least not the specific ingredients and flavour combinations.

As I am sure you can imagine, this chef really had the wind knocked out of his sails when the results came in. Why, he kept muttering darkly, did he bother putting so much effort into the creation of his dishes if his diners simply couldn't recall what they had eaten, nor the unusual flavour combinations that he had created? I told the chef not to be too hard on himself, that there was a psychological explanation for all of this and that it wasn't any reflection on his culinary skills. That the main thing to concentrate on was that people remembered their hedonic response, that they had really liked 'the experience'. And, if they also happened to engineer a memory of some confabulated meal, a construction of their overactive imaginations, well, so be it. It was especially surprising to hear from those diners who were convinced that their memories were so vivid they could almost taste the dish again. They were, as likely as not, imagining the flavour of something that they hadn't tasted, at least not at this chef's table! (All of this ought to make one wonder about the value of those online reviews.)

Using my most consoling tone, I told the chef that it was no use to try to fight the foibles of recollection. What he needed to do instead was to better understand the many ways in which memories fade and our minds deceive us. We mostly do not pay attention to what we taste. Rather, our brains just do a quality check first, to ensure that there is nothing wrong with the food or drink and that it tastes pretty much as we expected (or predicted) that it should. After that, once we know that we are safe, we devote our cognitive resources (what the psychologists call 'attention') to other, more interesting matters, like our dining companions, or what's on the TV or who has just sent us a text. That is, we no longer feel any need to concentrate on what we are consuming. And, as psychologists know only too well, unless you pay attention (e.g., to what you eat), you have little chance of remembering, even a few moments later, never mind after a few weeks or months. Emotion may also play a role here.

In fact, if you change the flavour of a food while eating (and yes, the experiment has been done), people often don't notice. It is as if we are all in a constant state of 'olfactory change blindness'. Intriguingly, this is something that the food companies have been trying to exploit to their, and hopefully our, advantage for a few years now. The basic idea is that you load all the tasty but unhealthy ingredients into the first and possibly last bite of a food, and reduce their concentration in the middle of the product, when the consumers are not paying so much attention to the tasting experience. Just think about a loaf of bread with the salt asymmetrically distributed towards the crust. The consumer will have a great-tasting first bite, and then their brain will 'fill in' the rest by assuming that it tastes exactly like the first mouthful did. This strategy will probably work just as long as the meal isn't high tea and the taster eating cucumber sandwiches with the crusts cut off! Or imagine something like a bar of chocolate, which most people will presumably start and finish at the ends, not in the middle. In fact, Unilever has a number of patents in just this space.

This innovative product development strategy is based, on the one hand, on the phenomenon of 'change blindness' and, on the other, on the assumption our brains make that things that appear the same probably taste the same too. The promise of the latest gastrophysics research is that by understanding such 'tricks of the mind', food and beverage companies, or at least the more innovative amongst them, will be able to deliver the same great taste that the consumer has come to expect without as much of those ingredients that we should *all* be consuming less of, like sugar, salt and fat.

Choice blindness

Do you think that you would notice the difference between two jams with a similar colour and texture, or two different flavours of tea? Most people will answer these questions in the affirmative. After all, isn't the very reason we buy one kind of jam rather than another, or keep multiple types of tea in our homes, precisely because we can differentiate their flavour? And yet gastrophysics research has highlighted some pretty worrying limitations concerning our perceptual abilities. It turns out that we actually have surprisingly little recollection (or awareness) of even that which we tasted only a few moments ago. In one classic demonstration of this phenomenon, known as 'choice blindness', shoppers (nearly 200 of them) in a Swedish supermarket were asked whether they would like to take part in a taste test. Those who agreed were then given two jams to evaluate. They were similar in terms of their colour and texture (e.g., blackcurrant versus blueberry). Once the shoppers had picked their favourite, they sampled it once again and said why they had chosen it, and what exactly made it so much nicer than the other jam. The shoppers were more than happy to oblige, regaling the experimenter with tales of how it was their favourite, or that it tasted especially good spread on toast, etc.

What many of the shoppers failed to notice, though, was that the jams had been switched before they tasted their 'preferred' spread the second time around. The experimenter was using double-ended jam jars in order to effect this switch unnoticed. In other words, the unsuspecting customers were justifying why they liked the spread that they had just rejected. Exactly the same thing happened in another experiment with fruit teas. Overall, the deception was noted by less than a third of shoppers. Even when the jams tasted very different – think cinnamon-apple and bitter grapefruit jam, or the sweet smell of mango versus the pungent aroma of aniseed-flavoured tea – only half of the switches were detected. What these findings imply is that many of the people tested actually did not have a clear memory of the flavour of the food that they had tasted only a moment or two earlier.

Such results, surprising though they undoubtedly are, nevertheless fit well with the research on blind taste tests. Consumers are convinced that they can pick out their preferred brand when given a range of alternatives to taste blind (i.e., without being able to see the labels). Time after time, they select one product and confidently assert that it is most definitely their preferred brand (presumably by comparing the taste to their memory of the taste). Why else, after all, would they be paying more for a branded product than for a cheaper unbranded or home-brand alternative? Only the thing is, in most cases the brand that they so confidently pick out happens to be something other than the brand that they normally choose. It is not that all of the products taste the same; more often than not they don't. It is just that our flavour memories aren't quite what we take them to be.

But surely this is not applicable to every product? Some of my colleagues always pipe up at this point that things are very different in the world of wine. They protest that I shouldn't believe the results of all those blind wine tasting studies in which people perform so badly, even the experts. And, truth be told, there certainly are some remarkable feats of blind wine tasting out there; that I don't doubt. But it is important here to distinguish between two alternative scenarios. On the one hand, there is the case of the expert who tries a wine blind and suddenly has a flashback to some long-ago vineyard where s/he first tasted it, remembering also who they were with and even what shoes they were wearing. In the alternative scenario, on the other hand, a more measured and rational assessment of a wine's sensory properties is undertaken by the taster. In the latter case, a careful process of elimination helps the expert to determine what the likely provenance of the wine is. Both can, of course, be most impressive, but only the former can really be said to demonstrate an outstanding feat of sensory flavour memory. It is interesting in this regard to see how guides to blind wine tasting tend to focus on the second approach. I suspect that, more often than not, the amazing feats of blind wine recognition are a matter of inference and cool calculation, rather than taste/flavour memory.

Have you heard of 'Sticktion'?

'In the context of experience management, [Sticktion] refers to a limited number of special clues that are sufficiently remarkable to be registered and remembered for some time, without being abrasive.

Sticktion stands out in the experience, but does not overpower it; well-designed, it is both memorable and related to the "motif" of the experience."[1] For all of you out there who are interested in knowing how to manage your customers' (or, for that matter, your friends') food and drink experiences, the good news is that there are various strategies that can be used to create 'stickier' memories, positive recollections that will hopefully form the basis of a decision to return to a given restaurant (or for those cooking at home, that will have your friends reminiscing about what a great cook you are). One suggestion is to deliver an unexpected gift, such as an amuse bouche, a little taster from the kitchen that the diners (or your guests) hadn't been anticipating, and certainly hadn't ordered. This is just the kind of positive surprise that is likely to stick in their memories long after they leave.

Similarly, the rise of the multi-course tasting menu also provides the opportunity to create stickier interactions, with the first taste of each and every dish providing a potential moment of 'flavour discovery'. Serving a large plate of the same food is, from the point of view of engineering great food memories, absolute madness. You know that people will remember the first few mouthfuls and that is it. This is the 'duration neglect' that we heard about earlier. The rest of the food will be lost to recollection no sooner than the plate has been cleared from the table. A wasted opportunity if ever there was one!

Anyone who is trying to design 'great-tasting memories' should probably also be thinking about primacy (and recency) effects here. Let me explain: if I were to give you a list of items to remember, like, say, the names of the dishes on a tasting menu, then you would be more likely to retain the first few courses and the last. The middle items will have to do more to stand out. No wonder, then, that so many chefs seem to really excel in their starters (not to mention their amuse bouche). Perhaps one can think of this as an intuitive example of experience engineering. If you know which dishes are most likely to end up sticking in people's memory, then working on really perfecting them can be time well spent, in terms of leaving your guests with the best impression (or memory) possible. Looking forward, it ought to be possible to figure out an ideal balance in terms of the number of courses that diners will have a reasonable chance of remembering while at the same time offering the chef enough scope to show what they are made of.

Just take the following as illustrative of the problems associated with trying to create more memorable food experiences. In one study, conducted in a restaurant, one-third of the diners questioned had no recollection whatsoever that they had eaten any bread, even minutes after having done so. This ought to make you think differently about how much effort you should put into that particular aspect of your dining experience. Indeed, it does seem like a growing number of restaurants in cities like New York no longer offer their diners bread. Intriguingly, it would also seem to argue against a primacy effect. But perhaps it is just that people simply don't think of the bread as a meaningful part of the meal. Perhaps it is treated as background, like the tablecloth, rather than the foreground experience of the dishes themselves.

Those experience engineers who have studied people's memories of mainstream restaurants find that our recollections rarely revolve around food. Just take the following as a case in point: when more than 120 customers who had eaten at a branch of the UK chain Pizza Hut were questioned a week after their visit, it turned out that the enthusiasm of the opening exchange with the restaurant staff – the warmth and energy with which the employees introduced themselves – was the single most salient memory for most of them. What was also important was how long it took to be acknowledged by a member of staff. Ultimately, if you know what your customers are most likely to remember, you are going to be in a better position to modify your food service offering, no matter whether you are a Michelin-starred restaurant or a gastropub. Knowledge really is power!

Do you remember what you ordered?

But, I hear you say, while the specifics of the tastes and flavours of the dishes that we have eaten previously might well fade from memory, don't we all at least remember our favourite dishes, or per-haps the chef's signature plate at the local eatery? For me, in my neighbourhood Italian, it is always fried whitebait followed by can-nelloni con carne. That is something that I *never* forget.* And whenever I find myself out for an Indian, then it's chicken jalfrezi, pilau rice and peshwari naan. The regularity with which I order exactly the same dishes is one of the things that convinces my wife that I am on the autistic end of the spectrum! But I would say it's not so much a question of neophobia (versus neophilia), but rather, if you know what you like, why change?

It is here that restaurants like The House of Wolf in Islington, north London, tend to fall down. Its business model was to offer a dining space for pop-up chefs and culinary artists, each doing a 4–6-week stint. Individually, each of the chefs who came into the kitchens was great. But they were also very different, one from the next. Hence, while you may be able to remember that you liked the chef who was cooking last time, you don't really have any specific positive memories of the dishes that might encourage your return, i.e., there is nothing concrete to look forward to. The management gurus are absolutely clear on this point: any really successful restaurant needs to have a few staple dishes that they are known for, dishes that customers remember and will come back to experience time and again. And indeed, The House of Wolf, just like many other restaurants with an ever-changing menu, didn't last long (the leaking roof probably didn't help much either). A shame, really, as I used to enjoy my role as professor-in-residence there. By contrast, think of those restaurant chains such as L'Entrecôte, where the menu is essentially fixed. That is precisely why the customer looks forward to returning – to eat exactly the same food that they remember so well from every previous visit. There are even those, like my wife's family, who have been dining at L'Entrecôte for more than four decades now, despite the perennially long queue, as you can't book a table (see also 'The Personalized Meal'). Presumably the queuing is also part of 'the experience', lending what appears to be some scarcity value.

---

* Though thinking about it carefully, maybe it is actually difficult to say whether I know what those dishes are generally supposed to taste like or whether I am remember-ing the specific taste of the dish the last time I went.

Do you remember what you ate?

Another technique that can help one's diners to create memorable dining experiences is to tell stories around the food. Eating at The Fat Duck restaurant is a good example. The experience starts with the diners being presented with a map and magnifying glass complete with duck's feet (see Figure 9.1). The idea that you are on some kind of journey has been ramped up, even while many of the dishes have stayed pretty much the same. Storytelling can, I think, help the diner to make sense of a multi-course tasting menu – a series of dishes that might otherwise seem like a random sequence from the chef's own personal hall of fame (a selection that might seem sentimental, even). By providing a storyline, a narrative framework, not only will the diner be better placed to parse or 'chunk' (that is, to group together items so that they can be processed or memorized more easily) the whole experience, but once 'chunked', the experience will be easier to remember too. This

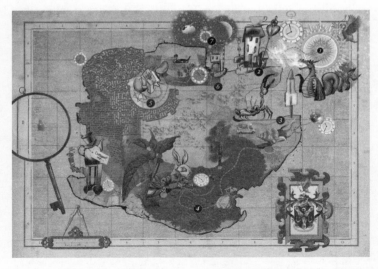

Figure 9.1. The map that diners are given to inspect (with the aid of a magnifying glass) on arrival at The Fat Duck.

becomes all the more important once we have lost the more traditional structure of the three- or five-course meal.

For special meals, it can also be a good idea to give the diner a copy of the menu to take away with them. The walls of the kitchen at home are adorned with framed examples of fabulous meals, like the menu from my first visit to Heston Blumenthal's restaurant in Bray. Even though it is now nearly fifteen years later, reading the descriptions of the dishes on the wall is all it takes to trigger so many pleasant memories (not necessarily of flavour, you understand, but rather recollections of the meal itself, and what I imagine the dishes could have tasted like). And, in this case, presumably, the more descriptive the menu, the better.

Even better, at least in terms of creating a lasting (or 'sticky') impression, was what happened when the menu was placed on the table midway through that meal. I still remember, just as vividly as if it was yesterday, picking up what looked like a regular vellum envelope, hoping to remind myself of the dishes that I had already polished off (and which I was already starting to forget – there had, after all, been so many) and to see what was coming next. I was shocked to feel something under my fingertips that had the texture of skin (the envelope had been specially treated). That was not at all what I had been expecting, and that moment of 'hidden surprise' has stayed with me ever since. Indeed, as noted earlier, more often than not it is the unusual or surprising experiences, the ones that capture our attention, that really force us to stop whatever else we may be doing (or concentrating on). It is the moments where we have to figure out what it is, exactly, that is going on that are remembered best. It is those events (and dishes) that need to be processed more thoroughly in order to be understood that will really stick in your memory. This is what the psychologists describe as 'depth of processing'; the deeper the processing, the better our recollection will be.

A final suggestion here, in terms of creating better food memories, relates to what is known as the 'end effect'. Our recollections of experiences tend to be dominated by what happened at the end, and food is no exception. Consequently, ending a meal on a high note can lead to greater remembered enjoyment. In one simple demonstration of this effect, researchers gave eighty people an oat cookie followed by a chocolate cookie. Another eighty people ate the cookies in the reverse order. Those who finished their snack with the more enjoyable chocolate cookie remembered the food as having tasted better when quizzed thirty minutes later. The end effect is presumably also what explains why 'all you can eat' meals are unlikely to be remembered too fondly. The end in such cases, at least in my personal experience (as a student, I hasten to add), is that feeling of being unpleasantly stuffed and knowing that you have eaten more than you should have. By contrast, those Italian restaurants where the meal ends with a surprise shot of limoncello might well be engineering a more positive memory, by ending the encounter with a mood-inducing unexpected gift. So, why not think about what you can do to surprise your dinner guests before they leave the table?

What is so special about mindful dining?

The next time you find yourself eating in front of the TV or computer, think carefully about what you are doing. As we have seen already, mindful (or attentive) eating and drinking is important, and anything that we can do to make ourselves more aware of what we are consuming is going to help in terms of increased enjoyment, enhanced delivery of multisensory stimulation, and quite possibly increased satiety too. But does mindful dining and drinking also lead to better memories of food and beverage experiences? It certainly feels like this ought to be the case, and that eating mindfully would also lead to reduced consumption subsequently (be it later in the meal or in subsequent meals). However, I think that we need to await further gastrophysics research in order to know for sure. That said, there is, I suppose, a sense in which the many diners taking all those pictures of their food and posting them on social media are creating an external memory of the event, and of the dishes themselves. They are an aide-memoire, if you will. It is those images, more than anything else, that will help people to remember what they probably know they will otherwise forget.

Away from the restaurant, though, one might ask what exactly we remember of our favourite foods? Consumers often complain whenever too drastic a change is introduced into the formulation of their favourite brands – the consumer backlash that was triggered by the introduction of New Coke or by the change in the shape of the Cadbury Dairy Milk bar, for example. Such behaviour could certainly be taken to imply that we must retain the taste of our favourite foods in memory. However, it turns out that our recollection of branded foods and beverages may only be as good as the last consumption episodes. This, after all, is what enables the food and beverage manufacturers to engage in their health-by-stealth strategies, gradually cutting the amounts of unhealthy ingredients, like sugar or salt, in their products without consumers ever realizing that anything has changed. Do it too fast and you'll have the consumer writing, calling and emailing in to complain that their favourite brands no longer taste the way they did! The challenge for the food companies is made all the more difficult, though, by the fact that even extrinsic product changes, such as to the colour of the can or the shape of the chocolate, can affect the perceived taste of the product (and give rise to an increase in consumer complaints), even if the formulation itself hasn't changed (see the 'Sight' chapter).

The gastrophysicist will tell you that as soon as we are alerted to a change or a potential modification (e.g., when we read 'low fat' or 'reduced salt' on the label) we start to pay more attention to what we are tasting. Thus it is normally a bad idea for a company to market a new reduced-fat, -sugar, or -salt product as such because that is likely to set expectations about the taste in the mind of the consumer (as discussed earlier), who will then pay more attention to the taste, looking for any differences. This is rarely a good thing. For as the Dutch sensory scientist Ep Köster notes, as far as the senses of smell, taste and mouthfeel are concerned, memory seems to be focused on detecting change rather than on identification and precise recognition of the food stimuli that we have encountered previously.

So, how are we to make sense of all this – failure to identify our favourite brands in all those blind taste tests on the one hand, and consumers' vociferous complaints when they perceive that their favourite brands have changed on the other? It doesn't seem to quite add up – or does it? Well, maybe we remain essentially blind to taste, i.e., we pay no attention unless our brains happen to detect that something is not quite as we expected it to be. Only then do we really start to concentrate. So if you are trying to nudge your family towards slightly healthier eating behaviours, the implications are clear: introduce changes to your cooking (such as reducing the salt) gradually, and whatever you do, don't let them know what you are up to.

The meal forgotten

I wouldn't want you to think that the 'meal remembered' is only of interest to those who, like in the movie *Total Recall*, want to insert particular memories into your mind (be it for financial or commercial gain). There is also an important element to this research, relevant to all of those unfortunate individuals who are losing or have totally lost their ability to recollect recent events, those who can forget, just as soon as the dishes have been cleared away from the table, that they have just polished off a three-course meal. For instance, amnesic patients suffering from Korsakoff's Syndrome (normally resulting from extreme alcoholism) may retain no memory whatsoever of having just eaten, and will happily start a second and even a third meal, if it is placed before them, providing they are momentarily distracted after the last meal has finished. One solution here is to leave visual cues of a recently finished meal lying around to provide an external reminder of the meal that has just been eaten.

In a related vein, a few years ago, I was a consultant on a project with a London-based scent expert and a design agency to help develop an intervention for early-stage Alzheimer's/dementia patients who forget to eat. The basic idea was that if these patients could be reminded to eat, they could maintain a semi-independent existence for a little longer than might otherwise be the case. The solution that my colleagues came up with is a plug-in that releases the scent of breakfast in the morning, of lunch at – you guessed it – lunchtime and enticing dinner aromas in the evening. The product has been available commercially for a few years now. In this case, one of the challenges for the gastrophysicist was to figure out which aromas to use, for while the smell of sizzling bacon might work as a breakfast cue for some, it clearly isn't an option for those who abide by certain religions. Furthermore, the foods we eat also change as the decades go by. So we had to try to find those food aromas that would be meaningful to individuals who were mostly expected to be of retirement age.

The importance of developing such sensory interventions becomes clear once you realize that close to 50 million people worldwide are currently suffering from dementia. In a small test of the designed solution, called 'Ode', fifty people living with dementia (along with their families) used the device for almost three months. The weight of more than half of those who took part either stabilized or increased, leading to an average weight gain of two kilograms. No wonder that it was voted the most innovative British Business Idea of 2013 by Small Business Cup.[2]

Hacking our food memories

There is also a really interesting line of research into the 'hacking' of food memories to bias people's food behaviours. For instance, researchers have shown that people's attitudes and behaviours around food can be subtly influenced simply by implanting false memories related to prior food-related experiences (e.g., telling someone that they once became ill after eating beetroot). Such misinformation, and the false memories that it can give rise to, can lead to significant behavioural change (e.g., lowered self-reported preference for and decreased consumption of beetroot). While the majority of the research conducted to date has been in the laboratory, there is growing interest in using such techniques to try to nudge people towards adopting healthier food behaviours. Could children, for instance, be encouraged to eat more vegetables by implanting false positive memories around previous pleasurable consumption episodes? And if they could, would it even be ethical to do so?

Remember, remember . . .

Ultimately, all that stays with us, once the meal is over, is our memory of it; we have only our recollections of the great occasions, and of the terrible ones too. The stuff in the middle – well, that is mostly just forgotten. Chefs, at least those with an eye on the future, would like to create food experiences that are more memorable, that have more 'sticktion', in the words of the experience engineers. Their long-term success, after all, depends upon it.

It is our recollection of the taste and flavour of foods that determines which restaurants we go back to, which food and beverage brands we stay loyal to, and even how much we decide to eat. In fact, simply reminding someone of what they ate earlier in the day (for lunch, say) can be enough to lead to a significant reduction in their subsequent snack consumption, when compared to those who were encouraged to recollect what they ate for lunch the previous day instead. Remembering what you have eaten recently can therefore be more important than you might have thought. Fortunately, the gastrophysics approach is increasingly assisting those who want you to recall more, as well as helping to improve the quality of life for those who can no longer remember.

And, although there isn't space to cover the topic in any depth here, consider too that the foods we eat can themselves also be used to evoke memories, as captured by the oft-cited case of Proust's madeleine. And just think of those so-called 'memory meals', like Thanksgiving in the US.

Let me end, though, with Jean Anthelme Brillat-Savarin's quote on taste and flavour in ageing, from his classic book *The Physiology of Taste*, published in 1825. The famous French gastronome writes: 'The pleasures of the table, belong to all times and all ages, to every country and to every day; they go hand in hand with all our other pleasures, outlast them, and remain to console us for their loss.'[3] As the old polymath knew only too well, eating and drinking constitute some of life's most enjoyable experiences. When our memory of those pleasures go, one might ask: what else is left?

10.   The Personalized Meal

You must have noticed how whenever you place an order in a branch of Starbucks the barista always asks for your name; then, when your beverage arrives, you find that it has been scrawled across the side of the cup. Now, this might seem to be necessary to avoid confusion at peak times, when there will be a large number of people standing expectantly at the counter, all waiting for their cappuccinos or skinny lattes. This is not merely a matter of operational convenience, though. Rather, this form of 'personalization' is company policy. Some even believe that it actually leads to a better experience for the customer. After all, one has the impression that the drink has been especially made for you. The question that the gastrophysicist really wants an answer to here, though, is whether this (or any other) form of personalization can make whatever you consume taste better too.

Everyone loves personalization

That personalization sells was amply demonstrated by the phenomenal success of the 'Share a Coke' offer in 2013 and 2014, whereby consumers could buy a bottle of the fizzy black stuff with their name printed on the label (see Figure 10.1). Make no mistake, this is superficial personalization (in the sense that the product itself hasn't changed). The drink is more or less the same the world over, and yet something about seeing your name emblazoned there on the front label changes the experience for you. So easy, so simple, and yet so incredibly effective: because of this campaign, sales increased for the first time in more than a decade.

No wonder, then, that many other food and beverage companies have been scrambling to try and copy Coke's success with their own examples of personalization. Indeed, according to an article that

Figure 10.1. During the summers of 2013 and 2014 Coke's marketing strategy made headlines when it swapped the powerful equity of its own brand for the names of consumers across seventy countries. The campaign had originally kicked off in Australia in 2011.

appeared in *Forbes Magazine*, 'Personalization is not a trend. It is a marketing tsunami.' Late in 2015, for instance, Moët & Chandon set up a number of photo booths in branches of Selfridges across the UK where their customers could upload a snap of themselves on to the front of their Mini Moët champagne bottles. The perfect Christmas present, apparently. Vedett has also been encouraging people to customize its beer bottles with their own pictures, and Frito-Lay did something very similar, offering 10,000 bags of potato chips for people to personalize with photos of their 'favourite summer moments'. In 2016, Kellogg's ran an offer whereby anyone who bought the requisite number of boxes of cereal could send off for their very own personalized spoon.*

---

* The writer Will Self managed to get a rather rude (or should that be puerile?) phrase printed on the front of his spoon; see W. Self, 'Finally my personalised spoon from Kellogg's® has turned up – and it's way better than I thought', *New Statesman*, 29

Have you heard of the 'self-prioritization effect'?

Why exactly should people respond differently to products that are associated with themselves? One possibility here relates to the 'self-prioritization effect'. Psychologists from Oxford recently discovered that arbitrary visual symbols (such as circles, squares and triangles), i.e., stimuli that have no intrinsic meaning, can nevertheless still take on a special significance just as soon as they become linked to us. In a typical study, one arbitrary stimulus (let's say a blue triangle) is associated with the self, while another is paired with a friend or someone else. The participants sitting in the laboratory are asked to press one button as rapidly as possible whenever they are shown the self-relevant object, and another whenever they see the stimulus belonging to the other person (perhaps a yellow square or a red circle). The results of a number of such studies have revealed that self-relevant objects are rapidly prioritized. That is, you see them sooner and respond to them faster than the other stimuli that have arbitrarily been classified as belonging to someone else. They have, in other words, become more salient because they are related or, in some sense, 'belong' to you.

My suspicion is that a similar phenomenon may well be at play when consumers come across a Styrofoam cup of coffee or bottle of Coke with their name on it. And presumably the birthday cake that comes to the table will also taste better to the person whose birthday it is, for much the same reason.

Do you have a favourite mug? For me, it is the orange one with a cartoon pig on one side and a chicken on the other that I look for every morning when I make myself a cappuccino. I get annoyed if I find that it is still in the dishwasher. Of course, the coffee is the same whichever cup I drink from. But somehow the experience feels different; the drink just doesn't taste the same. It could be

September 2015 (http://www.newstatesman.com/culture/food-drink/2015/09/finally-my-personalised-spoon-kellogg-s-has-turned-and-it-s-way-better-i).

that the self-prioritization effect might, in some small way, help to explain why it is that beverages always seem to taste better drunk from a favourite mug. One might think of this as a kind of 'sensation transference' (sometimes referred to as 'affective ventriloquism'), a concept first introduced more than half a century ago by the legendary North American marketer Louis Cheskin. This is where the feelings we have about the cup, our very own cup or mug (i.e., all those warm feelings of ownership and familiarity), are transferred to our perception of the contents. There is probably also a link here to the 'endowment effect': This favourite of the behavioural economists refers to the fact that we ascribe more value to things merely because we own them. This phenomenon is also known as the 'status quo bias'.

As far as I am aware, the proper experiment has yet to be conducted, but someone should really do it. All the budding gastrophysicist has to do is invite a group of people (thirty or forty is probably enough) to taste and evaluate coffee from their own cup and then coffee from someone else's mug (making sure, of course, to counterbalance the order of presentation). Maybe the coffee in the two cups is the same, or perhaps it's different – only the gastrophysicist knows for sure (or at least I hope they do). You almost don't have to do the study, though, in order to know which cup people would prefer to drink from. The funny thing is that when you quiz them people often feel almost embarrassed to admit they prefer their favourite mug, because at one level they believe it can't change the taste, that they are somehow just being silly. And yet, as a gastrophysicist, I firmly believe that this form of personalization really does make a difference to how much we enjoy the experience – a subtle one, perhaps, but significant nonetheless.

The 'cocktail party effect'

For many decades now, psychologists have been aware that your name seems to have a special meaning or significance: somehow it just 'pops out' of the background hubbub. You are probably most familiar with this happening when you suddenly become aware that someone else is talking about you while mingling at a noisy party, hence the name the 'cocktail party effect'. But, at one level, maybe this prioritization of your own name is not so surprising given just how much experience you have had of hearing it all your life. What is so noteworthy about self-prioritization, though, by contrast, is just how rapidly it comes to influence our behaviour. It occurs almost as soon as an object becomes, in some sense, 'ours'. Even a silly blue triangle is treated differently, once it has been assigned to us. What is more, and supporting these behavioural changes, brain-imaging studies show that different neural circuits are activated by self- as compared to other-relevant stimuli.

And that is not all: even the first letter of our last name has been shown to exert an influence over certain aspects of our behaviour. For instance, those whose surname starts with a letter that appears later in the alphabet are more likely to respond earlier in online auctions and limited-time offers, i.e., those with a surname beginning with the letter 'Z' are a little more impatient! This would appear to be tied to their names having been read out last at school, because married names don't exert anything like the same influence. There are a number of other intriguing phenomena, beyond this 'last-name effect'. You might find it surprising, for example, that we all tend to prefer items that have a similar spelling as the first letters of our own name (the 'name-letter effect'). Moreover, marketers know that we like those products, brands and even potential partners more if their name happens to share at least a few letters with our own.

Extending this line of reasoning to food might lead one to predict that we should all prefer, albeit ever so slightly, those dishes that share more letters with our own name as well. For me, *spicy* (*Spence*) food is the taste experience I always crave after a long trip. According to the research, the three shared letters with my surname may be doing at least some of the work here in terms of my liking for this sensation. Furthermore, whenever I order *chilli* con carne, another of my favourite dishes, I cannot stop myself from thinking about all the letters it shares with my own first name (*Charles*). Why not try this for yourself? How many of your favourite dishes share letters with your name? And, next time you meet a girl called Victoria, don't be surprised to find that she has something of a penchant for sponge cake.

Personalization in the restaurant

There are a number of ways in which the service offering can be personalized in the restaurant that are likely to enhance your meal experience. One relatively simple example here comes from those restaurants like Ricard Camarena's Arrop in Valencia, Spain, where the waiting staff take note of the type of bread that you choose when the basket is first brought to the table. When they return, they will very deliberately point to the bread that you selected last time and ask whether you would like the same again, or whether you would prefer something different. By this one act, they are letting you know, however subtly, that they are paying attention. Many other well-run restaurants use similar techniques.

Another thoughtful example here comes from a restaurant that we have heard a lot about in the previous chapters, namely the three-Michelin-starred Fat Duck. The waiting staff there carefully observe the diners at the start of the meal in order to figure out their handedness. A mental note is made of any lefties at the table and thereafter the dinner service is aligned accordingly. The interesting thing, though, is that no mention of this is made to those sitting at the table; in fact, all that the less-observant diner may be aware of is that the experience just seems to 'flow'. A few of the more attentive diners will probably spot the personalization, and hopefully appreciate the effort and attention to detail that has gone into creating a dining experience especially for them.

'Where everybody knows your name'

Who doesn't like to be recognized when returning to a favourite haunt? You know the sort of thing: 'Why, hello, Mr Spence, how nice to see you again.' This has been christened the 'Cheers' effect, after

Figure 10.2. Charlie Trotter waiting with his staff for the arrival of a VIP (in this case, Chicago mayor Rahm Emanuel). A welcome guaranteed to make the diner feel special!

the Boston bar in the 1980s sitcom. While it is unlikely that the staff in your local branch of Pizza Hut will remember your name,* the ultra-high-end restaurants take things to a whole new level when it comes to making their guests feel special. The ultimate example has to be Charlie Trotter's famous refrain of 'Kerbside!', which would ring out through the kitchen of his namesake restaurant in Chicago. According to London-based chef Jesse Dunford Wood (who worked there for a while), that was the cue that a VIP was about to arrive. Everyone in the kitchen would then march outside and line up in front of the restaurant's entrance to welcome the guest (see Figure 10.2). Those of you who remember the hit TV series *Downton Abbey* will recognize this scene; staff at the house welcome the lord of the manor like this when he returns home after a long trip.

New York uber-restaurateur Danny Meyer caused something of a stir back in 2010 with the publication of *Setting the Table*, his memoir

---

* Though, as we saw in 'The Meal Remembered', even at the lower end of the food chain, the warmth of the welcome turns out to be the single most memorable aspect of a restaurant visit likely determining whether you will return.

of a life in the restaurant business. Meyer has been in charge of a string of famous restaurants, including Union Square Cafe, Gramercy Tavern and Eleven Madison Park. Time and again in his book, he stresses the importance of personalizing the service at his restaurants. For years, they have been storing information about the diners when the initial booking is taken, so as to ensure a familiar welcome when the guests arrive. In fact, they keep a file of regular guests and their gastronomic peccadilloes. You can imagine the sort of thing: does so-and-so prefer to be seated in the window or hidden away in an alcove? What is their first name and, more importantly, do they like to be recognized or do they prefer to remain anonymous? Do they have a fondness for Super Tuscans, perhaps, or, like Jay-Z, prefer white burgundies . . . ?

While Meyer's New York restaurants are often cited as leading the way in terms of attentive and personalized service, many others claim to have adopted a similar policy. Famous Chicago venues such as Alinea, Next, Moto and iNG all try to find out something about those who are dining with them. According to Nick Kokonas, co-owner of Alinea, Next and The Aviary, they have kept a database of every single guest who has been on the premises since they opened. Initially, he says, the idea 'was simply to identify [guests] visually and thus greet them by name, like saying hello to an old friend you are greeting in your home'. Over time, though, this has morphed into the delivery of a more personalized experience for the diner. Even more surprising is the suggestion that the restaurateur might, on occasion, use such information to follow up on any of the regulars who haven't shown up in a while.[1]

How to make a first-time guest feel special

While it is easy to see how the restaurateur with a compendious Rolodex, or whatever the technological equivalent is these days, can make their regulars feel special, how can you give someone who has *never* been to one of your establishments the same experience?

Imagine for a moment how you would feel going to a restaurant in a new city where the doorman acknowledges you by name. Then you sit down, only to find that the waiter assigned to look after your table for the evening happens to come from your home town (somewhere far, far away). How freaky would that be? Don't worry, though, this is not ESP, rather just a sign that the restaurant has been googling you prior to your arrival. Justin Roller, for instance, the maître d' at Eleven Madison Park, is famous for googling every single one of the diners before they arrive, looking for anything that can help his staff to make diners feel both special and comfortable (i.e., almost as if they were at home). 'If, for example, Roller discovers it's a couple's anniversary, he'll then try to figure out which anniversary [. . .] All that googling pays off when the maître d' greets total strangers by name and wishes them a happy tenth anniversary before they've even taken off their coats. ("We want to evoke a sense of being welcomed home," [another staff member] says.)' It is hard not to notice how often commentators point to the outstanding customer service as an important part of the success of Meyer's restaurants. Now, at least you know the secret.[2]

Would it bother you if a restaurant googled you before you even walked through the door? Or would you welcome the practice because of the personalized service that you might receive as a result? Well, according to the results of a poll conducted back in 2010, nearly 40% of North Americans said that it was OK, their assumption being that it would lead to some kind of special treatment. A further 16% thought it was a little strange, but said that they could probably live with it. However, 15% of respondents thought it downright creepy. There is presumably a fine line to be drawn here between having a better experience as a result of personalization and feeling that one's privacy has somehow been violated. As one restaurant consultant interviewed by *The New York Times* put it: 'If you say, "I know you like a white Burgundy from the 1970s," that is creepy. Instead, you ask them what they like and point them in the direction of that white Burgundy.'[3]

When it slipped out in the press recently that The Fat Duck was googling its diners, several hundred reservations were immediately cancelled. Not a problem for a restaurant that is reputed to have 30,000 booking requests a day, but still a slight hiccup that one could have done without. The irony here is that the restaurant, like all those top North American venues, has actually been googling their guests for years. But that is beside the point: what is more interesting is the marked difference in reaction from North Americans. Perhaps the English are just that little bit more reserved.

What's next in terms of personalization?

The service philosophy in many of these high-end restaurants is clearly fundamental to their offering. The aim, at least at the very top, is for people to want to come back to the restaurant because of the service. Remember that poor service is the number-one criticism of diners year after year. A professional attitude is important, obviously, and good food helps too, but *personalization* is key. It is, after all, one of the best ways to make the diner feel special. The more personalized the service, the more likely we are to enjoy the experience, the better we will remember the food as tasting and the larger the tip we will leave (though this is apparently another area where the British tend to be a little more reserved than their North American counterparts).

The challenge, moving forward, at least as I see it, is how to take the idiosyncratic or one-off personalization of service at restaurants like Eleven Madison Park and industrialize it. For, ultimately, the canny restaurateur wants all of their diners to feel special, not just the lucky few who caught the maître d's eye when he googled them. Though, of course, when personalization becomes ubiquitous, it must surely lose some of its appeal. There is a very real danger of it coming across as artificially crafted as opposed to naturally friendly.

'Tell me when you were born and I will create a dish especially for you.' I can still remember a sentence to this effect appearing on the menu at The Fat Duck a decade or so ago, when the tasting menu was still optional. Nowadays, a more systematic approach to personalization has been adopted at the restaurant, one that revolves around nostalgia.* And rather than (or perhaps as well as) googling their diners, the restaurant staff ask about them directly. From the moment you manage to secure a booking (normally two months in advance), those working behind the scenes in Bray will be trying to find out some key information in order to personalize the experience table-side. The last time I went with my wife, we received a flurry of emails asking about our childhoods.

Part of the way in which such information is incorporated into the experience comes at the meal's end, when the miniature sweet shop is wheeled over to your table blowing cute little smoke rings from the chimney. This marvel of engineering looks like an ornate doll's house (and is rumoured to have cost more than a Rolls-Royce!). You are handed a coin and, when you insert it into the Sweet Shop (see Figure 10.3), drawers open and close in a seeming chaotic sequence. Eventually, though, the contraption† comes to what seems to be a haphazard stop, with one of the drawers left open. (It looks random, but, of course, it is not.) The waiter then hands the diner their very own bag of sweets from within the open drawer. The kinds of sweets that the diner finds in the bag should hopefully resonate with what they remember eating as a child, an update of 'the kid in a sweet shop' idea. Nostalgia is being used here to deliver a generic form of personalization (generic to those of a certain age, i.e., born in a specific decade). The hope is this interlude will help to trigger positive childhood memories and emotions that will come to colour the diner's recollection of the meal as a whole. The nostalgia/storytelling angle is still evolving.

While this level of personalization is currently restricted to

---

* In a lovely bit of publicity for the brand, it was the nostalgia element of a meal at The Duck that apparently saved one couple's marriage; see J. Tweedy, 'How dining at The Fat Duck saved a couple from divorce: Heston reveals warring lovers were reunited after eating "nostalgic" meal', *Daily Mail Online*, 16 December 2015 (http://www.dailymail.co.uk/femail/article-3362700/Heston-Blumenthal-says-dining-Fat-Duck-saved-couple-divorce.html).

† Though my favourite table automaton has to be the *Table Automaton featuring Diana and a Centaur*, created by Hans Jakob I. Backmann, in Augsburg (*c.*1602–6; https://artdone.wordpress.com/2016/05/10/celebration-125-years/hans-jakob-i-bachmann-table-automaton-featuring-diana-and-a-centaur-augsburg-ca-1602-06-khm-vienna/).

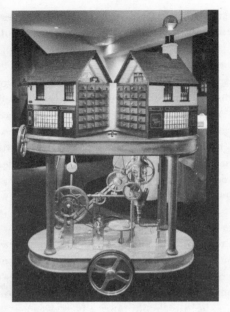

Figure 10.3. A personalized gift is waiting just for *you* inside one of the drawers of
the Sweet Shop at meal's end at The Fat Duck restaurant.

high-end restaurants, it is not going to stay that way for long. There
is already evidence that more mainstream venues are capitalizing on
the various online tools, such as Venga (and OpenTable), that allow
the restaurateur to pick up some useful 'diner-int.'. By integrating
such guest management and loyalty programmes with a restaurant's
point-of-sale systems, the staff can track their customers' average
spend, their favourite items from the menu and even their preferred
tipple. And – echoing what we saw earlier – some top restaurants use
it to record the handedness of their diners. The Venga system is not
cheap (currently coming in at somewhere around $149–$249 per
month per location), but it is a price that a growing number of restaurateurs feel it worth paying to give their guests that VIP treatment.
Here's a hint of what some are aspiring to: 'By the time a guest walks
through the front doors at Ping Pong Dim Sum in Washington D.C.,

marketing manager Myca Ferrer can already be fairly certain what he or she will order.'[4] Perhaps such predictive software could also play a role in helping to reduce the phenomenal amount of food waste seen at most restaurants today.

For those of you thinking about how to personalize your own dinner party, why not start by placing a name card where you want people to sit. You could use this crowd-control measure to keep all the bores together, as the Queen's party planner advises. But who knows, it might make your guests enjoy their meal a little more too. And of course you could try googling anyone you don't know so well. It can't do any harm, after all – well, just as long as they either don't care or else don't find out!

At the chef's table

This concept ticks a number of the right boxes in terms of on-trend contemporary eating practices. A small number of diners get to sit around a central space behind which the chef prepares, or at the very least finishes, the dishes. It gives the solo diner something to look at, not to mention someone to talk to. It also enables diners to see their food being freshly prepared. And depending on the chef, there may well be an element of theatre and spectacle too. Crucially, there is also more scope for personalization. Typically, the format is a *prix fixe* ('fixed price') menu, as at the three-Michelin-starred chef's table at Brooklyn Fare, in New York, and 12 Chairs, in Shanghai. It is hard not to be personal, after all, when the diner can look directly into the chef's eyes.

The ultimate in terms of personalized dining is the private chef. While this is something that you would normally expect to see only in the homes of the rich and famous, you can nevertheless find a few restaurants out there offering pretty much this level of service. Fuad's, in Houston, is one such venue. The head chef, Joseph Mashkoori, will come to your table and ask you what you'd like to eat. He might proffer a few suggestions, but he will insist that he is happy to cook whatever you want – from chateaubriand to a Philly steak sandwich. Meanwhile, in New York, Jehangir Mehta does something similar in his culinary venture called 'Me and You'. According to the website, the diner is promised: 'A unique private dining experience like none other. Every dish is custom crafted to suit your taste, play to your palate and inspire your juices.' The Solo Per Due restaurant in Vacone, Italy, also offers a very private meal experience, with only a single table for two.

A matter of choice

At one level, this drive towards the personalization of the dining experience would seem to be at odds with the simultaneous rise of the tasting menu (where the diners are given virtually no choice about what they will eat).* The waiter may ask the diner about any allergies and dietary requirements they have, but that's pretty much it. In fact, most of the time the only decision that the diner has to worry about is whether or not to go for the wine pairing, assuming that there is one. But what, exactly, explains the growing popularity of the tasting menu? Isn't this, in some sense, the very antithesis of personalization?

Some commentators have wanted to link it to attempts by chefs and restaurateurs to increase the memorability of the meal. As we saw in 'The Meal Remembered', the more dishes the diner tries, the more opportunity there is for 'Sticktion'; tasting menus typically have more courses than one would choose if ordering from the normal à la carte menu. It is easy to imagine how, when everyone is enjoying the same food at the same time, it is going to enhance the feeling that one really is *sharing* a meal (see 'Social Dining'). And then, from the chef's perspective, perhaps the economies associated with serving only a limited range of seasonal produce, and being able to select one's best offerings, also helps make up for the lack of choice.† There are, of

---

* What is worse, at many of those restaurants where the tasting menu is still optional, they normally insist that either *everyone* at the table has it or else *no one* does.

† One could, perhaps, also frame it as the chef trying to wrest back some semblance of control, given the explosion of dietary requirements, allergies and intolerances that

course, those of a more cynical persuasion who see this as just another way to extract more cash from the diner, since tasting menus tend to command a higher price too. A more positive take, though, might simply be to say that diners don't like the process of choosing (and thus having to decide against all the things that they can't have). Perhaps there is also a link to the unwritten rule in the restaurant business that the better the venue, the smaller the number of options the diner is given.[5]

Some people are offended by the loss of choice. According to Tim Hayward, writing in the *Financial Times*: 'Menus without choice blaspheme against the doctrine of dining.'[6] However, it is not only the tasting menu where the diners' range of options is limited. From à la carte to those joints serving only a single dish, it can feel as if there is a steady reduction of choice across many styles of dining. In a sense, though, this is all just an extension of the *prix-fixe* menu or even the table d'hôte that used to be such a common feature of dining out in France and some other European countries. Talking of which, one incredibly successful and long-running restaurant with limited choice is the L'Entrecôte chain in France (and London, New York, Bogotá and beyond). While there is a menu for drinks and desserts, there is only one option as far as the starter and main course is concerned: Salad, followed by steak (the diner is given the choice only of how they would like it cooked), a delicious sauce (the recipe is kept secret) and French fries *à volonté*. Virtually no options, no personalization, and yet people queue, sometimes for more than an hour, to get a table (they do not take reservations). So, you have to ask yourself, how much choice do diners really want?

Well, on the one hand, it is certainly true that what the marketers used to tell us – namely, that more choice is *always* better – no longer seems to hold true (if it ever did). Give people too much choice and they can feel overloaded. If you *are* going to give diners choice, seven seems to be the magic number: seven starters, seven to ten mains, and seven desserts. Any less and there is a danger of there being too little

---

diners present with nowadays. When I see what friends who are chefs have to deal with, I sometimes wonder how they manage at all.

choice. Any more and the diner may find it difficult to decide. Of course, the trick for those restaurants wishing to offer more choices than such a fixed format allows is to break the menu up into a number of sections. How many? You guessed it: the recommendation is again seven.

Rory Sutherland, Vice Chairman of the Ogilvy & Mather Group, UK, has an anecdote that fits nicely here about how airlines ended up selling far more cheap tickets as soon as they started to restrict the number of discount destinations they offered. Again, this would seem to run counter to sound economic principles. Surely, the more options there are, the more likely it is that a customer will find a destination they want. And yet the sales data emphatically shows the opposite to be true. Behavioural economists know only too well that we really can be paralysed by too much choice.[7] Presumably, that is why we have recently started to see the emergence of the 'condiment sommelier' in places like New York, whose job is to guide you through the options for, say, mustard or mayo when the range of alternatives becomes too formidable.

The 'Ikea effect'

You know the situation: you are at home preparing a meal for your friends and you think that you have really excelled yourself this time, that the food tastes absolutely fabulous. Your guests, polite as always, tell you that the food is delicious. But what do they *really* think? My advice as a gastrophysicist is don't trust what they say; rather, watch what they do. The question still remains: are your guests just being polite, or do they perhaps taste the food differently because they didn't make it themselves?

Marketers have a name for the increased value that things seem to have if we make them ourselves: they call it the 'Ikea effect'. In other words, just because you assemble that wooden table yourself, it is worth more to you than if it had come pre-built. But while there is plenty of evidence that co-creation invests value in the outcome as far as flimsy furniture is concerned, what we want to know here is whether the same is true of a meal that you make for your friends. And does the answer depend on whether you are cooking from scratch versus cooking from a meal kit or partially prepared meal?

Norwegian researchers have started to address these questions. In one series of experiments, they had various groups of individuals (not all students, you will be relieved to hear) make a meal from a kit in a kitchen laboratory. The researchers assessed what people said after they had prepared it themselves, or after being told that someone else had cooked it. Intriguingly, those who had prepared the meal themselves (or, to put it better, who thought that they had) rated it as tasting much better than those who believed that they were evaluating a dish made by someone else. This despite the fact that everyone was actually trying the same food (an Indian tikka masala dinner, if you must know). Furthermore, those who had to fry the meat and prepare the food as prescribed on the side of the package rated it as tasting better than those who merely had to stir and heat. In other words, the more involved the cook was in the act of creation, the better the end result tasted (at least to them).

So the chances are that if you make your friends dinner the food really will taste a little different to you, and this difference is likely to be more pronounced if you cooked from scratch (rather than 'cheating' with a pre-prepared meal). However, the bad news is that, if anything, the food probably tastes *better* to you than it does to anyone else (since they didn't make it). What all this means in practice is that you should get your friends involved in the kitchen, so that the food will taste better to them too.

Making cake

There is an interesting link here to one of the marketers' favourite case studies, involving the Betty Crocker cake mix. According to the oft-recounted tale, this powdered cake mix failed on its launch in the marketplace back in the middle of the last century. The product's fortunes took off only once a certain marketing executive[8] had figured out that the product formula should be changed so that the home cook had to add an egg to the mix. This obviously increased the effort involved in baking the cake, something that any rational analysis would surely say was a bad idea. And yet sales steadily climbed. The suggestion was that by adding the egg the cook somehow became invested in the process – that is, they could feel more like they were actually cooking! And it is quite possible that the finished cake would actually taste better to whoever made it too, because of their greater involvement in its making – the Ikea effect again.

You find the Betty Crocker story recounted all over the place – even the top North American food writer Michael Pollan talks about it in one of his bestselling books. It sounds almost too good a story to be true, right? Well, that is most probably because it is, at least according to a 2013 history of cake mixes that appeared in *Bon Appétit* magazine. It turns out that a patent for 'a cake mix that required the home baker to add a fresh egg', rather than a powdered one, was granted back in 1935 to a company called P. Duff & Sons. It was in the 1950s that sales of cake mix (both with and without fresh eggs) stalled. The innovation that actually revived the fortunes of the cake mix was the introduction not of the egg but rather icing; that is, the explosion of interest in *customizing* one's cakes and buns with fancy designs was what really reinvigorated sales. So the Betty Crocker story is just a load of old baloney; nevertheless, the point about the importance of personalization still stands.

Before we leave this topic, I am going to have to pick a fight. And not just with anyone, but with Nobel Prize-winning experimental psychologist and now behavioural economist Daniel Kahneman. He is in print, in *The New York Times*, no less, trying to justify the claim that 'sandwiches taste better when someone else makes them'. This would certainly appear to be an idea that resonates with journalists around the world, as the story has been picked up by any number of news outlets. Nevertheless, if you trace the story back to its sources, it turns out that the assertion is based on nothing more than speculation. In other words, no one has, at least as far as I can tell, ever done the proper sandwich study. And, given what we have just learnt about the Ikea effect, I see little reason to believe that we would, in fact, prefer other people's sandwiches to our own. I don't know about you, but I certainly think my own sandwiches taste pretty darn good! And I am not alone on this one; from what people are saying on the online discussion forums it would seem that many others share my intuition. Once again, then, another important piece of gastrophysics research waiting to be done.

I would like to end this chapter by asking: why is it OK to *customize* some dishes but not others? *Customization* can be thought of as a form of personalization, but one where control remains firmly in the hands of the consumer (or customer). It should feel empowering (rather than creepy). It was the opportunities for customization that were provided by the encouragement to ice one's cake that led to the revival in sales of cake mix. In the case of dining, customization occurs when the customer has some say over how their choice of dish is prepared, seasoned and/or served, everything from how spicy their curry through to how well done their burger. When the waiter offers to grate some Parmesan over your pasta at your local Italian, or you instinctively reach for the salt or pepper next time you go to the steakhouse – these are all examples of customization. But when should we feel entitled to customize our food? Let me illustrate with an infamous tale from the annals of restaurant folklore.

'Would you be so kind as to pass the salt and pepper?'

Marco Pierre White, the first British celebrity chef to be awarded three Michelin stars, comes from my home town of Leeds, in the north of England. His cookbook *White Heat* was one of the first I used when learning to cook, given to me by my sister on the occasion of my sixteenth birthday. It is still on (and often off) the shelf three decades later, containing, as it does, a truly sublime recipe for lemon tart. However, this chef first came to fame – or should that be public notoriety? – for throwing diners out of his restaurant should they be impertinent enough to ask for the salt and pepper. White said that it was an insult for any diner to want to season (or *customize*, we might say) their food in *his* restaurant. Seasoning a dish is, after all, the chef's remit, no? So, if a diner were to ask for the salt and pepper, then they could mean nothing but to insult the chef. Why so? Well, because they would be implying that the kitchens hadn't done their job properly. That, at least, is how the chef saw it.

In hindsight, one can view this incident as an early sign of the ascendance of the star (or *prima donna*) chef, no longer content to be hidden away in some dark and hot back room, never to be seen and rarely, if ever, to be acknowledged by the dining public. It presages, I would say, today's situation, whereby the celebrity chef has an open kitchen at the front of the restaurant for *everyone* to see. The chef, in other words, is now very much the star of the show. They call the shots, and don't we all know it!

However, while it was Marco Pierre White's tantrum that first brought the question of customization to the table, as it were, it does seem to me that the salt and pepper have been slowly becoming a little more elusive in restaurants up and down the land. You certainly don't find them in the majority of high-end modernist establishments.* So, one might ask, is it not just choice that the chefs wish to restrict but also the diner's opportunity to customize their food? (At the same time, of course, the top restaurateurs are ramping up the level of personalization in other parts of the meal experience.)

So why do we customize some dishes but not others?

Come to think of it, I would never dream of asking for salt and pepper in The Fat Duck. But why not, I ask myself. (And why don't we have sugar and citric acid shakers to season our desserts too, I wonder.) It is partly about trusting the skills of the chef, or rather the culinary team beavering away in the kitchens. However, it partly relates to the nature of the food that is being served. Much of it is so unlike anything that I (or you, I presume) have ever had before that it can be hard to know quite what the chef had in mind, what they were

---

* My suspicion is that the pretensions of the chef/kitchen and the presence (or lack thereof) of the salt and pepper are most probably correlated.

trying to achieve. As such, I have no internal standard against which to judge the dishes. All I know for sure is that they are delicious. So, given that I do not really know what I am *supposed* to be tasting, I am really not sure what outcome I would be aiming for, were I to try seasoning it myself.

We can contrast this with the situation when you go out for a steak. Not just any steak, mind you; let's imagine you've just ordered the £140 Japanese wagyu 8oz rib-eye steak from star Austrian chef Wolfgang Puck's restaurant Cut, in The Dorchester Hotel on Park Lane, London. In this case, I most definitely *do* expect to be asked how I would like my meat to be cooked. And there would be a riot if I didn't find the salt and pepper close at hand!* Notice that we are not so far off the price point of The Fat Duck (at least not once you've ordered your fries, sides and starter, etc.), and yet customization of the food is expected in one case and not even worth contemplating in the other. So price, or the skill of the kitchen, can't be the whole story here.

So what, then, *is* the difference? Well, I have had many steaks before, so I have some sort of internal standard that I am aiming for. My food memories may, of course, be inaccurate (see 'The Meal Remembered') but still I have an idea (or at least think I do) about the taste. Whenever a pepper grinder is available, though, I will habitually reach for it and add a liberal dose to my food, even before I have taken the first mouthful. How are we to explain such behaviour? Could it be that, by this simple act, I am somehow making the dish *mine* (customizing it, if you will), and that this somehow automatically makes the food taste better (in much the same way that it might when we rotate the plate just a little once the waiter has placed it down before us). Alternatively, however, I suppose it is also possible that we might all realize that our own taste preferences tend to reside at one end of the taste spectrum (for instance, I know that I like my food spicier than most other people), and hence that most food that

---

* Marco Pierre White now has a chain of steakhouses; we even have one of White's restaurants here in Oxford. I was tempted to go along and see what would happen if I asked for the salt and pepper myself, but it turns out that White actually has nothing to do with many of the restaurants that bear his name and image; they are just franchises.

has been prepared for the masses will taste better to us if we adapt it to our own personal profile. That said, perhaps the real distinction here relates to the degree of adulteration of the raw ingredients on their way to becoming a dish on the menu, and thereafter on my plate. So for relatively unprocessed foods, like steak, it is OK to season; it may even be expected. But as soon as that meat has been cooked in or presented with a sauce, say, then the justification for seasoning the dish ourselves is diminished somewhat, especially when we are in the hands of a top chef. And when it comes to the fabulous concoctions on the menu at The Fat Duck, they are so highly processed one may no longer have any real sense of quite what the raw ingredients were; that is, they have been transformed into something totally new. Under such conditions, then, there is simply less of an incentive for customization, as the goal of doing so is no longer so clear. Note that it is not that personalization is removed entirely in such venues, rather that it occurs in other aspects of the service or the meal.

My personal answer

So, in closing, let's return to the question of whether chef Marco Pierre White was right when he said that the diners had no right to customize his food. Is the seasoning of a dish really something that is best left to the chef (assuming that they have got the requisite number of Michelin stars)? Ultimately, isn't the diner always right? And never forget, as we saw in the 'Taste' chapter, we all live in very different taste worlds. In fact, once you realize just how distinct and personal your taste, smell and even flavour perception really are, there can be no looking back. So, in the future, will the foods and drinks served in restaurants themselves start to match our own personal taste profiles? This was another of those far-reaching ideas that was prefigured by the Futurists. The movement's founder, F. T. Marinetti, wrote that '[w]e will create meals rich in different qualities, in which for each person dishes will be designed which take into account sex, character, profession, and sensibility'. Recently, consumers have been offered the opportunity to personalize the taste of everything from chocolate

(Maison Cailler) through to champagne (Duval-Leroy), while Illy has developed a new system to allow their customers to adjust the sensory profile of their coffee.[9]

So, given all of the above, I am sure you'll understand why I recommend that you leave the salt and pepper on the table the next time you throw a dinner party, no matter how good (you think) your cooking is, and no matter what you end up doing to the ingredients. As a gastrophysicist, I'd say that the fact that the diner chooses to season their dish shouldn't be thought of as an insult to the chef, but rather as a form of customization that recognizes the very different taste worlds in which we all live.

11. The Experiential Meal

'Did you enjoy the experience?' This, the question that was asked time and again by the various service staff who ushered the guests through the different stages of Albert Adrià's 'About 50 days', held in London's Café Royal. But since when did this become *the* question? Why not ask whether the diners enjoyed the food instead? Getting to the bottom of this matter is the topic of this chapter. It is the story of the ubiquitous rise of the 'experience industries', first predicted by Alvin Toffler in his bestselling 1970 book *Future Shock*.

B. Joseph Pine, II, and James H. Gilmore, building on Philip Kotler's early (1974) ideas about atmospherics deserve the credit for introducing the 'experience economy' to the marketplace. Their point was simply that consumers are not really buying meals and drinks, or, for that matter, any other kind of product or service; rather, what people want is to enjoy, and increasingly to share, 'experiences'. These experiences are, by definition, multisensory. The realization, then, that dining out isn't really about fulfilling any kind of nutritional need helps make sense of why so many people now want to know the answer to Adrià's question.

Look around the world of fine dining and you find a growing number of chefs and restaurateurs promising to deliver multisensory dining experiences. For example, chef Andoni Aduriz (now chef at Mugaritz, in San Sebastián) had this to say of his time working at elBulli under the guidance of Ferran Adrià: 'For him, what mattered most was the experience, what one felt when eating at elBulli. Everything necessary would be done to create this experience.' Or take the following newspaper description of one of Marco Pierre White's franchise restaurants: 'The steakhouse, opened two years ago, is described on its website as being "all about the experience; the buzz, the atmosphere and enjoying the company of friends and family in a gorgeous comfortable surrounding." '[1]

On the rise of theatrical dining

As we will see in this chapter, restaurant cuisine is moving from its traditional function of providing nutrition – or restoration, as in the original meaning of the term 'restaurant' – towards becoming a medium for artistic expression. Restaurants are becoming stages; the waiters and chefs at some of the world's top establishments are increasingly playing the role of actors and magicians. First there was atmospherics, then came the theatre, the storytelling and the magic at mealtimes: This really is the very heart and soul of 'off the plate' dining. It is all the rage amongst the chefs that you find jostling for position in the San Pellegrino list of the world's top fifty restaurants. Some have been tempted to suggest, though, that such lists may be exerting undue influence . . . 'At Eleven Madison Park [in New York . . .] Daniel Humm and Will Guidara devised an entire program of changes motivated in part by the perception that "the San Pellegrino voters reward restaurants with a strong sense of place, and of theatre." They included a three-card-monte dessert and – further belaboring the locavore trend – a cheese-and-beer course that emerged from an old-fashioned Central Park picnic basket.' Of course, not everyone is happy. As one commentator puts it: 'If the wine industry has become Parkerized, then the restaurant world might be said to have been Pellegrinoed.' Meanwhile, David Chang (of Momofuku fame) describes the archetypal '50 Best' restaurant thus: 'It's a Chinese restaurant by a guy who worked for Adrià, Redzepi, and Keller. He cooks over fire. Everything is a story of his terroir. He has his own farm and hand-dives for his own sea urchins.'[2] In fact, some have suggested that knowing just how important it is to sell 'the experience' may actually be what's driving many of the contemporary trends in the world of high-end dining and drinking.

Sublimotion, in Ibiza, currently offers the world's most expensive restaurant meal. Should you be lucky enough to book a seat, you will need to find close to 1,500 euros a head. At that price, the twenty-course tasting menu can't just be about the food, now, can it? The meal must be fabulous, that's a given. But there needs to be so much more. The underlying assumption is that diners are willing to pay a hefty premium for an *experience* rather than for a mere meal, no matter how tasty the food.

The rise of the open kitchen, not to mention the growing popularity of the 'at the chef's table' dining concept, can be framed in terms of their ability to transform the very preparation of food into a kind of theatre. In fact, the tour of the kitchen is becoming an increasingly common component of a meal at many high-end restaurants. Juliet Kinsman, writing in the *Independent on Sunday* had the following to say: 'If you'd told proprietors then that one day diners would demand to eyeball the cooking team at work, they'd have blanched. Now butchery's brought to the fore and labour is part of the flavour – we want display as we dine. At ABaC Restaurant & Hotel in Barcelona, part of the thrill is being directed to your table (where you'll savour Jordi Cruz's two-Michelin-star 14- or 21-course menus) through the 200sq metre kitchen. And at the Typing Room in east London, I'm sure the fact that I could see Lee Wescott piping chestnut cream into the bowl made his dishes all the more delicious.'[3]

Even the removal of choice that we came across in 'The Personalized Meal' can be seen through the lens of experience design. But this, let me tell you, is really only just the beginning. There is so much more that can, and increasingly is, being done. According to one textbook on restaurant design published in 2011, restaurants are more than 50% theatre (see Figure 11.1). From where I am standing, that percentage would appear to be increasing year on year.

Have you come across theatrical plating?

There has been a growing tendency in recent years to turn the serving of food into a theatrical stunt too, this time by plating directly on to the table itself. At Alinea, in Chicago, for instance, a number of the desserts require a performance that can last for several minutes. For one dish, the waiters first lay a waterproof tablecloth over the table, before bringing all manner of sauces and ingredients. Next, one of the chefs emerges from the kitchen and starts to 'plate the table' in front of

Figure 11.1. Chef Jesse Dunford Wood sabre-ing the top off a bottle of sparkling wine. A most theatrical start to a meal at the chef's table at Parlour, in north London.

the amazed diners by breaking up the solid elements and painting with the sauces (both drop by drop and 'Jackson Pollocking' the table-top). Given all the practice that they have undoubtedly had, they manage to paint the dessert on to the table with great skill. Something similar happens at Sublimotion, in Shanghai, where the 'waiting staff appear with palettes of ingredients and "paint" an edible version of Gustav Klimt's "The Kiss" on the table'. Once the chefs have finished their work, the diners tuck in, eating straight from the table.

Meanwhile, a little closer to home, one of the chefs I have worked with is Jesse Dunford Wood, who is famous for theatrically plating the dessert directly on to the chef's table (hidden away in a walled alcove between the kitchens and the rest of the restaurant) at Parlour, in Kensal Rise, north London. The chef wields a dangerous weapon: the blow torch! Headphones are handed out to every diner, playing tracks that have been specially chosen to be familiar and hopefully to trigger an emotional response. When I was last there, the iconic theme from *2001 A Space Odyssey* was up first, followed by Gene Wilder singing something from *Willy Wonka and the Chocolate Factory*. Scented smoke was pumped out from a hole in the wall – this was, in other words, a truly multisensory experience. This example of performance around the plating of a dish is particularly interesting given that the chef himself listens to the same soundtracks as the diners while he works. Everyone at the table is connected in a private, but shared, sonic experience.

There is, of course, always a danger that things can be taken too far: for example, Dive, the Los Angeles submarine-themed restaurant from Steven Spielberg. According to those who have been, the lighting was pretty extreme. There was a wall of monitors along one wall constantly flickering away, showing submarine-themed movie clips. One commentator described what would happen inside: 'Periodically, all lighting is extinguished except for intense red lights that whir and flash while a loudspeaker barks "Dive! Dive!" '[4] Sounds pretty intense, right? Perhaps a little too arousing? No wonder the restaurant closed down.

One way to deliver a memorable dining experience is by hosting it in a most dramatic or unusual location: venues such as the underwater restaurant in the Maldives or the Dinner in the Sky concept spring to mind (see Figure 11.2). A somewhat less extreme, though no less successful, version of the latter concept went by the name of The Electrolux Cube. For a while, this transparent structure was situated on top of the Royal Festival Hall, on London's South Bank, and a stream of Michelin-starred British chefs popped up there in order to serve the eighteen guests. When you add in a great view, the experience is undoubtedly elevated beyond your average pop-up dining event. The cube has now appeared at a host of scenic locations around Europe: a rooftop overlooking the Piazza del Duomo in Milan, in Stockholm (atop the Swedish capital's Royal Opera House). It has been to Brussels too. Part of the success of this venture can presumably be put down to the fact that this is a limited offering (scarcity being highly valued in the restaurant sector at the moment).

'The everything else'

How would you like to go to a restaurant where the atmosphere changed from one course to the next? The top chefs with money to burn can do this using technology, others (on much smaller budgets)

Figure 11.2. *Top:* The underwater Ithaa restaurant opened in April 2005, seating up to fourteen diners and located five metres below sea level at the Conrad Maldives Rangali Island. *Bottom:* 'Dinner in the sky': these diners are having a unique experience suspended tens of metres above the ground. This is more about the experience than the food, one presumes.

manage to achieve much the same effect by having their diners move from room to room as they switch between courses. As chef Grant Achatz put it (when contemplating overhauling the experience at his restaurant Alinea), 'perhaps diners will do a portion the tasting menu in one space, before moving to another environment that is completely different in its configuration, design elements, lighting, even aroma.'[5] Make no mistake about it: today, we are increasingly seeing a shift (often facilitated by technology) towards a much more dynamic and adventurous approach to the dining experience, one that involves storytelling, added theatre – oh, and possibly a dash of magic too. So, welcome to the all-new world of experiential dining. And here we are not just talking about playing with the colour of the lighting, or synchronizing the music or soundscapes to match each and every one of the dishes. We have already come across a number of examples of ambient aroma being used to complement specific dishes (see 'Smell'). Some chefs, like Paco Roncero over in Ibiza's Hard Rock Hotel, have gone further: they are even playing with the atmosphere (i.e., the temperature and humidity) of the dining rooms they control.

The aim of the next generation of experience providers is to enhance, that is, to complement, what is hopefully already a fabulous product offering (not to distract you from a poor one), by optimizing 'the everything else'. It is the chefs who are really at the top of their game (those with two or three Michelin stars, and who appear in the San Pellegrino World's 50 Best Restaurants list year in, year out) who are innovating here; a number of them have realized that no matter how good the food they put on the plate, unless they are in control of 'the everything else', they really can't hope to optimize the experience for their diners. Of course, as was noted earlier, one could turn things around and suggest that these chefs are focusing on 'off the plate' dining precisely because they know that this is what the San Pellegrino judges are looking for. As one commentator put it: 'Chefs play to the list, mindful of its aesthetic preferences and its methodological weaknesses.'[6] The chefs are absolutely clear on this point: French chef Paul Pairet, of Ultraviolet in Shanghai, insists that the goal of changing the multisensory atmosphere on a course-by-course basis is to 'intensify the focus on the food, not distract from it'.[7]

Elsewhere, he has been quoted as saying: 'You can't escape from what I'm trying to convey. Everything will lead you to [develop] a strong focus on the dish.'[8] The technology-enabled atmospheric projections on the walls and tables of some of these futuristic dining rooms undoubtedly allow for more theatre and storytelling, elements that are key when it comes to trying to hold the diners' attention/interest over what may well be a 15–20+-course tasting menu.

If you want to know what to expect, here's a description by one journalist who was lucky enough to dine at Pairet's Ultraviolet: 'Dinner starts dramatically with an apple wasabi sorbet, frozen and cut into wafers. A Gothic abbey appears on the walls, the air is filled with holy incense and AC/DC's "Hells Bells" assaults the ears.'[9] Meanwhile, an evening at Sublimotion has been described as: 'an emotional "theatre of the senses" . . . a night of gastronomy, mixology and technology'.[10] Ultraviolet bills itself as the first of its kind to bring together the latest in technology to create a fully immersive multisensory dining experience. It opened in May 2012 to a maelstrom of interest from the world's media.*

Other no-holds-barred gastronomic events include the one-off Gelinaz dinners. The food at these events is whipped-up by some of the world's top chefs and the courses are interspersed with music, dance, magic and video. Just as well, I suppose, given that these events may last for anything up to eight hours. The El Celler de Can Roca restaurant in Spain has regularly been voted top of the World's 50 Best Restaurants list in recent years. Back in May 2013, the chefs (the Roca brothers) worked with music director Zubin Mehta and visual artist Franc Aleu to create a fabulous twelve-course culinary opera called 'El Somni'. This one-off dinner was held for twelve carefully selected guests (talk about exclusive) in a specially designed rotunda in Barcelona. A once-in-a-lifetime experience – quite literally! An

---

* The dates here are important in terms of assessing precedence: Ultraviolet opened in 2012, Sublimotion in 2014. And while a number of today's chefs may well fight over who deserves acknowledgement for precedence, really it was the Futurists who got there first (see 'Back to the Futurists'). One might wonder whether restaurants like these should be considered under the umbrella of 'eatertainment'; I suspect not, given the pejorative associations of the latter term.

amazing sound system was installed especially for the event and visually stunning projections surrounded the diners – no expense spared. Indeed, I dread to think how much it must have cost to put this dining event on; there is no way that it could break even, even if the diners were paying through the nose (which they weren't). My guess, though, is that it was probably worth it for the many brands sponsoring this venture, given the huge amount of international publicity it attracted.

Performance at the table

Do you think that you'd enjoy your dessert more if it came out from the kitchens at the same time as a cellist sat down next to you and played a specially composed piece of music, or even just a sustained musical chord? It would, at the very least, be a rather unique experience, would it not?[11] Composing music especially for mealtimes certainly isn't new; go back far enough (to the mid sixteenth century) and one finds _Tafelmusik_ (literally 'table-music') being composed for feasts and other special dining occasions. Now, composers, artists and sonic designers are once again taking up the challenge of designing music specifically for the meal. While once the music was, in some sense, composed for the _occasion_, now it is designed to match the _food itself_.

Some of you may be wondering just what influence the atmospheric soundscapes and music being played in a growing number of restaurants have on your experience? You might wonder about its effect on the taste of the food, not to mention on how much you enjoy the total experience. Note that we are not just talking about fancy restaurant meals but also the local gastropub. In an earlier chapter, we saw how the sound of the sea could be used to enhance the taste of oysters. Subsequent research by the Condiment Junkie multisensory experiential design team has shown that playing the sounds of the English summertime enhances the perceived fruitiness and freshness of strawberries. Put this evidence together with the literature on sonic seasoning and it is clear that both sensory-discriminative (i.e., what it is) and hedonic ratings (how much you like it) of the food are likely to be affected by what diners hear. All the more reason, then, to try and get it right.

One interesting challenge that crops up once you start to think about designing music, or soundscapes, to complement a particular dish (or even an entire meal) is that the structure and duration of each track probably has to be quite different from that of traditional music. In fact, music that has been especially composed for the meal (or dish) probably has more in common with the kind of sonic backdrop that people design for video games, say, than a top-forty hit. Ideally, it should be a little bit repetitive, consistent over time, but with the potential to evolve seamlessly as the diner moves from one course (or level) to the next. This is exactly what the sound designer Ben Houge aims for in his innovative sonic installations. Back in 2012, for instance, Houge worked with chef Jason Bond on a series of dinners in Bond's restaurant Bondir in Cambridge, Massachusetts. Each table was outfitted with one loudspeaker for each diner, which resulted in a total of thirty channels of coordinated, real-time, algorithmic, spatially deployed sound that was designed to work even if different tables of diners arrived at different times.

Storytelling at the table

According to an article that appeared in *The New York Times* back in 2012: 'Restaurants in the very top echelon these days – Noma in Copenhagen, Alinea in Chicago, Mugaritz and Arzak in Spain – sell cooking as a sort of abstract art or experimental storytelling.'[12] An excellent example is the Alice in Wonderland theme that runs through a number of the courses at The Fat Duck – the 'Mad Hatter's Tea Party' dish comes straight out of the pages of Lewis Carroll. When the restaurant reopened late in 2015 (after refurbishment), Blumenthal turned to Lee Hall, the writer of the film *Billy Elliot*, to help weave the menu into a story, which means that 'the menu will now be a story. It will have an introduction and a number of chapters and the chapter headings that will give you an idea of what is coming.' The top chef hasn't stopped there, though. He has also been talking to the press about trying to redefine the very nature of the restaurant. The emphasis on narrative has been ramped up substantially: 'The fact is The Fat Duck is about storytelling. I wanted to think about the whole approach of what we do in those terms.'[13] All of this explains chef Jozef Youssef's decision to mount the details of the dishes that he served in his new 'Gastrophysics' dining concept in the books placed innocuously on the table. Meanwhile, at Alinea, chef Grant Achatz has been wondering what would change if dinner at his restaurant were to be like the set of a play.

Magic is increasingly making an appearance at the dining table. At Eleven Madison Park in New York, for instance, they have been looking at introducing a card trick as part of the dessert service. Blumenthal conferred with magicians while experimenting with a flaming sorbet that would ignite at the click of the waiter's fingers. According to one journalist: 'Blumenthal created it with a magician so, at the click of a waiter's fingers, the barley sorbet in a bowl of hidden compartments bursts alight, turning warm outside yet remaining ice-cold inside. As a fire crackles around the sorbet, a rolling vapour of whisky and leather transports you to some Scottish hunting lodge at Christmas.'[14] Unbelievably, the bowls are rumoured to cost £1,000 a piece.

Theatre at the table

Why does it always have to be dinner and a show? Why not simply combine the two and dine *while* watching the show, or where dinner *is*, in some sense, *the show*? The dining events offered by Madeleine's Madteater in Copenhagen are often described as free-form experimental food theatre. As one journalist described it: 'It's art you experience with all five senses, the most satisfying performance in town. Madteater is precisely as its name translates: food theater. [. . .] We were transforming the act of eating into exactly that: an act. I felt equal parts diner, performer and audience member in a restaurant that channelled all at once the opera, an art gallery and a shrink's office. It was strange. It was delicious.'[15]

Figure 11.3. The Tickets Bar in Barcelona, a recent collaboration between the brothers Ferran and Albert Adrià.

What exactly would you think you were looking at were you to walk past the shopfront shown in Figure 11.3? It looks like some kind of theatre, right? But this is actually a tapas bar! According to one description: 'The atmosphere draws equal inspiration from the theater and the circus, with a nod to Willy Wonka and his overstimulating chocolate factory. It's a place where chefs toil away at various workstations, waiters prance around like theater ushers and morsels of food arrive with the flair and mystery of a vaudeville act.' So, as dining becomes more theatrical, more entertaining, it becomes natural to make the restaurant *look* like a theatre too.

Not only that but one might also consider the idea of selling tickets for the show. In fact, this is exactly what American chef Grant Achatz decided to do with his Chicago restaurant Next. Anyone wanting to eat there can simply buy a ticket in advance from the website. And just like the theatre (and with airlines), cheaper seats are offered for off-peak shows/meals. So a seat at Monday lunchtime will cost you less than a prime-time seat on a Saturday evening. It's an intriguing concept. No surprise, then, that a number of other restaurants and pop-up dining events have subsequently adopted a similar model; on the Ultraviolet website, for instance, you are encouraged to 'book your seats now.'

In the years to come, we are going to see a continued blurring of the boundary between theatrical and culinary experience. Take the highly innovative Punchdrunk Theatre company. Still incredibly vivid in my memory, as I am sure in many others', is 'Sleep no more', the multi-story retelling of Shakespeare's *Macbeth* in an abandoned warehouse block in New York City. This was an immersive theatre experience like no other. So, as actors, singers and magicians increasingly make their way into the dining rooms, the question arises: what would you get if you mashed up something like Punchdrunk Theatre with multisensory dining/drinking? Well, funny you should mention that, because the people behind Punchdrunk opened a restaurant. According to founder Felix Barrett, they originally developed a distinct narrative for the restaurant involving a cast of twelve actors. However, when the concept was trialled, the feeling was that ' "people weren't ready to watch theater" while they ate. Expense, he suggested, was also a factor. So now there are fewer, less formal theatrical accompaniments to the catering.'[16]

Perhaps it is because I happen to be married to a Colombian that I think so, but there really is nothing quite like Andrés Carne de Res. This restaurant on the outskirts of Bogotá has actors, musicians, magicians and a host of other performers wandering haphazardly between the tables in an atmospheric higgledy-piggledy assortment of wooden shacks. Sorry, that is the best I can do to describe it.* You really need to experience it for yourself. Best go in the evening though, when, after the food has been served, the tables turn into an impromptu dance floor. Taking people through different spaces, rather than using technology to create different atmospheres within the same space, can be a low-tech and, more importantly, lower-cost solution to delivering experiential dining. After all, not everyone

---

* In 2016, this restaurant was rightly voted one of South America's top 50; see http://www.theworlds50best.com/latinamerica/en/The-List/41-50/Andres-Carne-de-Res.html. And for those of you who are worrying, the country is now much safer than all those scare-mongering TV shows like to make out.

has the budget or technical support offered to star chefs like Paul Pairet, Paco Roncero or the three brothers Roca! Reducing the expense associated with delivering the multisensory experience offers the opportunity to provide something that becomes a little more scaleable.

One example of the low-tech approach comes from Gingerline's 'Chambers of Flavour' experience in London. Small groups of diners enjoy a four- or five-course meal, each course being served in a different room with a contrasting theatrical experience. According to Suz Mountford, founder of this immersive dining enterprise: 'Guests book having no idea of what to expect and can journey through anything from enchanted forests to a spaceship to a sunset beach, meeting all sorts of crazy characters along the way . . . We've always wanted to firmly cast the dining experience as a creative space to stimulate not just the taste buds but all of the senses.' Actors, dancers, and performers are again all part of the experience.[17]

Amongst all this talk of spectacle at the table, I would be remiss not to mention the truly spectacular meal held in February 1783 in Paris. Alexandre Balthazar Laurent Grimod de la Reynière, son of a wealthy tax farmer and nephew of one of Louis XVI's ministers, hosted a dinner in which hundreds of spectators watched the proceedings from a gallery – converting hospitality into some kind of theatrical show. Invitations to the meal took the form of ornate burial announcements. Just take the following description:

> Like a banquet of freemasons, to which contemporaries compared it, Grimod's supper made heavy use of arcane ritual and semi-democratic pretensions [. . .] Grimod's guests had to pass through an entrance hall and a series of chambers before reaching a darkened waiting room, and, finally, the inner sanctum of the dining room. In one room, heralds dressed in Roman robes examined the guests' invitations; in the next, an armed and helmeted 'strange and terrifying monk' subjected them to further scrutiny; before a note-taking fellow in lawyer's garb greeted and interrogated the twenty-two guests [. . .] in the final stage of initiation, two hired hands dressed as choir boys perfumed the guests with incense.

This spectacular dinner was so far ahead of its time that it deserves to be remembered to this day. One could even frame it as a piece of performance art with food at the centre – this, from almost 250 years ago!

Performance art with food

Looking back over the last half-century or so, one finds many examples of performance artists incorporating food, the preparation of a dish and/or its consumption into their work. The idea goes right back to the usual suspects, the Futurists, who 'aimed to marry art and gastronomy, to transform dining into a type of performance art'.[18] But there are many more recent exponents in this space; just take, for instance, Alison Knowles, an American experimental artist who served up a salad for 300 people to eat while listening to Mozart at the Tate Modern in London (see Figure 11.4 for one such 'happening'). This participatory event, *Make a Salad*, grew out of the artists' collective Fluxus movement back in the 1960s. This piece plays on the notion of consistency – a theme that we will come back to in the final chapter. As the artist herself puts it: 'The salad will be made again for

Figure 11.4. *Make a Salad* (1962–) by Alison Knowles, a participatory performance art piece with food (normally for several hundred people).

several hundred spectators [. . .] Beginning the event, a Mozart duo for violin and cello is followed by production of the salad by the artist and eating of the salad by the audience. The salad is always different as Mozart remains the same.'[19]

Nothing, though, could ever match the harrowing ordeal faced by the sixteen guests who attended Barbara Smith's six-course *Ritual Meal* (1969). This performative event started with the invitees waiting outside someone's home for an hour. They were told repeatedly by a voice over a Tannoy to: 'Please wait, please wait.' On being let in, the guests were immersed in a space that was filled with the loud, pulsing sound of a beating heart. Videos of open-heart surgery were projected on the walls and ceiling. If that sounds bad enough, wait till you hear what happened next:

> Eight waiters (four men wearing surgical scrubs and masks and four women wearing masks, black tights, and leotards) led them to a table. Prior to entering the house, the guests had to put on surgical scrubs [. . .] The guests were then served a meal like they had never seen before. In keeping with the surgical 'theme,' the eating utensils were surgical instruments. Meat had to be cut with scalpels. Wine, served in test tubes, resembled blood or urine. In this charged atmosphere, ordinary food took on extraordinary connotations, an effect that Smith enhanced by the preparation and presentation of the food. Pureed fruit was served in plasma bottles. Raw food, such as eggs and chicken livers, that had to be cooked at the table were included in the dinner along with plates of cottage cheese embedded with a small pepper resembling an organ. Although the food was actually quite good, the dining experience was intensely uncomfortable for the guests, who couldn't put down their wine/test tubes and were sometimes forced to eat with their hands.[20]

You can get an idea of what it must have been like from a close-up of the hands of one of the guests at the performance (see Figure 11.5).

Is food art? Traditionally the answer has been very definitely no, the key distinction being that the viewer/diner was not 'disinterested', in Wittgenstein's terminology. And yet it is clear that many chefs are increasingly taking their inspiration from the world of art,

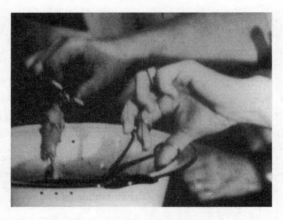

Figure 11.5. Barbara Smith's *Ritual Meal* (1969). Still feeling hungry?

some even referring to themselves as 'artists'. Certainly, as we move away from the quaint notion that dining out has only to do with nutrition and sustenance, we may even see the emergence of artistic dishes that do not necessarily taste that good. Something like this has already started to happen! Just take one of the latest menus at the Mugaritz restaurant, in San Sebastián. There's a dish (only one, mind) that the chef knows may be difficult for the diner to enjoy (a local delicacy of dried fish) and yet which plays an important part in the story of the meal that is served in this rural spot. The dish stays on the menu despite the negative comments of some diners online. As chef Andoni explains, in his book: 'There was a key moment in the evolution of Mugaritz when we realized that we were serving certain things that, objectively, weren't "good", but which had great emotional power. The dish "Roasted and Raw Vegetables, Wild and Cultivated Shoots and Leaves", for example, is eaten in a sort of altered state of consciousness [. . .] Plant bitterness is difficult to overcome, and this dish, out of context, could even be branded unpleasant. It is undoubtedly a proposition that creates mild discomfort.'

Getting to the bottom of whether food can ever be considered as an art form, and whether chefs (at least the best of them) should really be thought of as artists is a debate that is likely to rumble on and on.

It is certainly not something that I have any realistic hope of resolving here, and certainly not in a paragraph or two. My guess, for what it is worth, is that it will become increasingly natural to consider the top chefs as artists. And at some point in the not too distant future, we will all come to wonder why we ever didn't!

On the future of experiential dining

Any of you who find it difficult to imagine the experience of dining out being any different in the future from the way it has been in the past, just remember, the restaurant as we know it is actually a fairly modern invention, coming into existence in the latter half of the seventeenth and early eighteenth century in Paris. Perhaps it's time for the format to be updated. A less radical way to think about this, though, is perhaps to consider the relative distribution (and pricing) of different kinds of restaurant changing. Researchers who have ana-lysed the situation distinguish between the *fête spéciale*, 'where dining has been elevated to an event of extraordinary stature', the amusement restaurant and the convenience restaurant.[21] The rise of multisensory experiential dining is likely to enlarge the first two categories at the cost of the latter.

Rest assured – or be very afraid, depending on your perspective. Before too long, many of our more mundane everyday dining and drinking experiences, be they in chain restaurants, hotels, food and wine shops, at home, or perhaps even in the air, will be accompanied by some sort of multisensory experiential cues. The hope (at least for me) is that these atmospheric stimuli will have been scientifically designed by one of the growing number of gastrophysicists out there to modulate, and preferably to enhance, some aspect (or aspects) of your tasting experience. Given what we know about the limits of the diner's brain, gastrophysicists interested in multisensory experience design are well placed to help to deliver experiences that are both stimulating and memorable without being overpowering.

There is also growing interest in moving from 'eatertainment' to 'edutainment', exemplified by Heston's Dinner in London. At this restaurant, the stories behind each of the dishes, relating to British food history, are retold. A similar dose of storytelling was also a part of the chef Jozef Youssef's 'Mexico' concept from Kitchen Theory. Take, for example, 'The Venison Dance' dish, which was preceded by a short video of the dance performed by the State Ballet of Mexico. The 'Memories of Oaxaca' dish served as part of the same concept also had a video to help set the scene for diners.

In closing, it is important to stress the key challenge faced by theme restaurants. Given that diners know what to expect the second time around, those who wish to stage 'experiences' have to work constantly to refresh them. As the successful New York restaurateur Danny Meyer notes: 'Showmanship can be a tricky pursuit [. . . because] the more theatrical it gets, there does come a point when you just can't see that play another time.'[22] On the flipside (see 'The Meal Remembered'), though – there is also the comfort that comes with knowing exactly what we are going to eat.

12. Digital Dining

Would your cocktail taste as nice or your dinner as delicious if you found out that it had been prepared by a robot? Would you trust a robot chef to season your food for you? Really? And how would you feel about being served by a robot waiter? Sounds like science fiction, right? Well, these things are already becoming reality, albeit in only a small number of venues so far. Like it or not, a whole host of digital technologies have become ever more closely intertwined with our everyday experience of food and drink. There are already digital menus out there that will send your order straight to the bar or kitchen. Pizza Hut even trialled a 'subconscious' menu that could, or so it was claimed, magically read your mind and tell you the three toppings that you wanted on your pizza, without you having to say a word. The menu tracked the customers' eye movements while they viewed the digitally displayed options. No worries, though, if you happen not to like what your subconscious chose, as you could always stare at the restart button and begin the process all over again! I have to say, though, that this example smacks of marketing gimmickry rather than a serious attempt to envision the restaurant of the future. Elsewhere, though, one comes across headlines talking of a '*Star Trek* "replicator" that can recreate ANY meal in 30 seconds'.

Soon there may no longer be any need to reach for the sugar. At least, not if you know about the latest findings from the field of sonic seasoning, which can be dispensed by the mobile devices that so many of us now carry around with us. In fact, digital artefacts will probably become a ubiquitous part of our multisensory food and drink experiences in the near future. It is more than likely that we will first come across these new technologies at the tables of some of the world's top modernist restaurants or cocktail bars. But from there, it is only a few short steps to their introduction in chain restaurants and the home environment.* Furthermore, many global food and beverage brands are eager to get a slice of the action here too. So let's jump straight in and see what tomorrow's digital dinners might look like, and who – or should that be *what*? – exactly might be making them.

Would you like a 3D food printer?

Read the newspapers and you'd be hard pressed not to come away with the belief that the 3D food printer is going to be the all-new must-have gadget in the home kitchen. If you haven't heard about the Foodini, the Bocusini or the (rather more mundanely titled) 3D Systems ChefJet Pro 3D, then you are probably not an 'it' cook. Chefs are now using 3D printers to impress diners by creating foods the likes of which they have never seen before (see Figure 12.1 for one beautiful example of 3D food printing). So will the 3D food printer really become the microwave of the future? It is obviously in the best interests of the manufacturers to try and convince you that it is, but I beg to differ. Don't get me wrong, it's not that I see *no* use for them, it's just that my best guess is that they will end up in the kitchens of a few fancy modernist restaurants and in food innovation centres, not in the home kitchen (one of the few exceptions to my earlier predictions that what starts in the modernist kitchen and restaurant will trickle down and eventually find itself in the local restaurant or even the home.)[1]

One chef who has apparently become enamoured with the possibilities offered by 3D food printing is Paco Perez. He uses one at his restaurant La Enoteca, at the Hotel Arts in Barcelona, to create food forms that could not otherwise be made (intricate architectural renditions of famous buildings, for instance). I was also intrigued by the recent release of an inkjet printer that allows the barista to create amazing latte art (e.g., showing realistic portraits of famous people). And until his untimely death a few years ago, Homaru Cantu was

---

* See Buster Keaton's short comedy film *The Electric House* (1922) for an early take on what the digital dining room of the future might look like.

Figure 12.1. 'Caesar's Flower of Life': seasoned bread in 3D-printed sacred-geometric 'Flower of Life' design, with assorted flowers and vegetables. An eclectic eight-course dinner was 3D-printed and served using byFlow Focus 3D-Printers, prepared with fresh natural ingredients and borrowing innovative multisensory techniques from the world of molecular gastronomy.

famous for printing edible menus for his restaurant Moto, in Chicago. The inventive chef managed this trick by hacking a regular printer. In May 2013, NASA offered him a six-month Phase 1 study contract to develop printing technologies capable of combining shelf-stable macronutrients, micronutrients and flavours to produce personalized food products for long-duration space missions. Cantu generated lots of excitement when he started talking to the press, especially when he mentioned the future of printing 3D pizzas in space. When the story broke, though, NASA was outraged. A number of those working in the area felt that all the press coverage largely demeaned (and distracted from) the serious science underpinning the development of adequate food supplies for long-distance space flight. No wonder, then, that funding for Cantu's project was not extended to the next stage.

Currently, commercial food printers are still a little too expensive for widespread home uptake, coming in at somewhere around $1,000 apiece. In time, that figure will probably come down, as it always does with technology. But even if they were giving them away, I still wonder who would be daft enough to get one for their home kitchen. Should you be wondering why I have become so curmudgeonly all of a sudden, well, you need to ask yourself how long it actually takes to print each perfectly formed morsel of food. My guess is that you would probably have to start preparing your very own unique pasta shapes a good few days in advance if you were thinking of inviting all your friends over for dinner. So while you could make it yourself with your new kitchen toy, don't forget we still have those quaint things called shops. And if you do find yourself craving a printer, just ask yourself who, exactly, is going to clean the tubes out each time after you have used your glistening new kitchen appliance. You might also wonder how much your electricity bill would go up. So, I ask you once again, is it really worth it?

3D-printed foods have novelty value, yes. But, beyond that, what new food experiences could they possibly give you? Is there a unique selling point here? What is it, exactly, that this machine allows you to do that you couldn't do otherwise? And even if – for some unknown reason – sales of 3D food printers do eventually pick up, my guess is that you will find all these sorry devices stored away lonely and unloved, collecting dust at the back of the cupboard, within a couple of years of the hype reaching its apex. That is to say, they will end up in exactly the same position as so many of those home breadmakers and food processors – the *must-have* kitchen appliances of recent decades.*

However, I could imagine creating the perfectly shaped chocolate, one that snugly fit the contours of the average human tongue and could therefore deliver a more intense flavour hit than any one of those seemingly accidentally shaped confections currently out there in the marketplace. Just imagine a chocolate that stimulated all of your taste buds simultaneously. But as soon as the perfect shape had

---

* That said, wait a few decades more, and they are likely to cycle back into fashion once again; K. Mansey, 'Gadgets of the 1970s get their fizz back', *The Sunday Times*, 15 December 2013, p. 23.

been identified (assuming, of course, that there is one), then the industrial production lines could be reconfigured to create it on a massive scale.

So, if you believe me when I say that we are not going to be eating 3D-printed dinners at home any time soon, then where else might we first experience the intrusion of digital technology into our food and beverage experiences? Well, many of you will have come across the next one already, namely the digital menu.

Do you like ordering from a digital menu?

No, me neither, though one does increasingly find digital menus in trendy high-end bars and restaurants. Now, this ought to make sense, right, at least on paper? No more worries that your waiter will forget something when they stubbornly refuse to write anything down when taking the orders. Digital menus should also allow for any changes to the vintages of the wines to be updated on the list in pretty much real time. This would at least deal with one of my pet peeves – when restaurants have one vintage on the wine list and then bring out a (generally inferior) younger bottle because they have run out of the good one, often without telling you (and expecting you to be happy to pay the same price for it). And, in theory, the digital menu should enable the restaurateur or bar owner to incorporate any seasonal dishes or drinks on to their menu, thus allowing them to bin the blackboard – you know, the one with the daily specials chalked all over it.

And yet, if you are anything like me, you will have realized that it just doesn't feel right. Maybe it is because I am not a millennial, but I have to say that when someone insists that I order from a digital menu, the dining/drinking experience is somehow diminished. Why so? Well, there are a couple of reasons: on the one hand, it is important, I think, to remember that dining is a fundamentally *social* activity (see 'Social Dining'). Part of the reason that we go to a restaurant or bar in the first place is for the interaction with the staff. Danny Meyer, one of the most successful restaurateurs of our time, nailed it, I think, when he said: 'Despite high-tech enhancements, restaurants will always remain a hands-on, high-touch, people-oriented business. Nothing will ever replace shaking people's hands, smiling, and looking them in the eye as a genuine means of welcoming them. And that is why hospitality – unlike widgets – is not something you can stamp out on an assembly line.'[2]

There are no two ways about this: digital menus take away from that social exchange and make for a much more transactional kind of experience. Some would say that placing an order from a digital menu is just cold. I, for one, am glad to see that restaurateurs and bar owners are slowly coming to their senses and getting rid of their digital menus. And not before time either, if you ask me. The only place where they make sense, I think, is in those venues where all that anyone wants is a rapid and efficient transaction, say, when grabbing a quick bite to eat at an airport.*

The other problem is that most digital menus look pretty much identical to the printed version. Why? Surely, going digital should be an opportunity to do something radically different. If a modernist chef were to present you with one, someone like Grant Achatz in Chicago, or Juan Maria Arzak in Spain, you *know* that it wouldn't simply be a digital replica of a paper menu. One of the few interesting examples is the digital menu that one sees on the table at Inamo, an Asian fusion restaurant in London. Not only can you order simply by touching the relevant item projected on to the table's surface but you can also see what the various dishes look like before ordering them.† The digital nature of the interaction also means that presenting pictures of the dishes doesn't seem anything like as tacky as it would if exactly the same images were to be displayed on a regular printed menu. There is, then, a tangible benefit here for the diner from 'going

---

* Intriguingly, though, the one place where you do find such impersonal service quite regularly is in Japan, where, on entering many restaurants/noodle bars, you find a picture menu from which you can order, and then wait for someone to place the food in front of you.

† In a way, this is again reminiscent of Japan, where many restaurants have a selection of hyper-realistic plastic meals on display at the entrance, showing what the food options look like. Indeed, I have just such a magnetic sushi set proudly displayed on the filing cabinets in my office.

digital'. You can even order a taxi home directly from the table-top! Of course, given the growing interest in where the food we eat comes from, one could also imagine being shown information about the history of the ingredients, which farm they came from, etc.

Digital menus also offer the possibility of delivering a more curated meal. Take Mother in Stockholm, where the menus are also projected directly on to the table-top, and the diners are quizzed about which foods they prefer. Then, a number of recommendations are made for dishes that the diner might like. (The only missed opportunity here is the fact that the system does not keep your details from previous visits – a trick of personalization that, as we have seen, is likely to make a meal both more enjoyable and more memorable.)

One other interesting use of digital interactive menus comes from The Weeny Weaning restaurant opened by Ella's Kitchen in 2014. This, the world's first sensory restaurant for babies, has been designed to encourage healthy eating from an early age, or so the blurb goes. According to one report: 'Little ones will be seated in highchairs at interactive tables, from which they will be able to choose from their very own digital menu, allowing them to order their own mains and desserts [. . .] Depending on the number of times they tap a particular food icon over a 30-second period, the digital menu responds accordingly and the waiters bring the children their selected choice of food.'[3] The next generation are likely to be much more open to this kind of digital interface with food, no matter whether or not they were exposed to it as a baby.

Tasting the tablet

Why would you use a real plate when you can use a tablet instead? (Or should I have said that the other way around?) This is, though, one of the ways in which technology has already started to change our visual experience in the modernist restaurant; there are chefs who serve some of their food from tablet computers rather than plates. (Who knows, perhaps this is the perfect way to make use of all those surplus tablets sitting around in all the restaurants and bars where they have figured out that digital menus are a waste of time!) Chef Andreas Caminada recently served one of the courses at his Swiss restaurant on top of a tablet displaying the image of a round white plate: an ironic take on digital plating.

A few years ago, we were playing around with the idea of serving seafood from a tablet (see Figure 12.2).[4] The diner would catch a glimpse of the sun glinting off the waves and the sand on the seashore, so realistic that they could almost touch it. That was the hope anyway! Combine the sight of the seashore with the sounds of the sea (about which more below), and the seafood may well taste better. At the very least, plating from a tablet should offer the creative chef increased freedom as far as the storytelling around a dish is concerned. While so far this is happening only in a few cutting-edge restaurants, one could imagine, in the future, all of us repurposing our own tablet computers at mealtimes.

Figure 12.2. How long before high-end restaurants start serving food from a tablet rather than a round white plate? Top Spanish chef Elena Arzak says: 'At Arzak in San Sebastián, certain dishes are served over a digital tablet: grilled lemons with shrimp and patchouli sit atop a fired-up grill with the noise of crackling flames. [. . .] We experimented with serving the dish on and off the tablet and diners always said that having the image and the sound intensified the flavours of the dish and made it even more enjoyable. We're keen to use new technology to further augment the meal.'[5]

Some of you out there will be appalled: why on earth would I spend good money on a tablet computer only to eat off the damn thing? you may be muttering darkly to yourself. What, exactly, is so wrong with the good old-fashioned round white plate? Has the professor actually lost his marbles? Don't get me wrong, I am certainly not saying that the tablet is going to be the ideal option for every kind of food. Even I can't imagine that it would be much fun trying to eat a big juicy steak, say, from a tablet. Probably best to keep serving this one on a wooden board. Canapés and other finger foods might be more the thing, at least till you get the hang of this all-new digitally enhanced plateware.

In my defence, though, let me at least point out that some tablets are waterproof so you could, I presume, put them in the dishwasher straight after use, should the need arise. (Perhaps the professor has lost his marbles after all!) I can easily foresee how serving food off a tablet could also provide the ideal means of ensuring the perfect colour contrast between your food and the plateware (or, in this case, the tablet screen) against which it is viewed (see the 'Sight' chapter). Ultimately, though, I think that eating from a tablet will go mainstream only if the diner's experience of the dishes served in the modernist restaurant is radically altered, i.e., enhanced, by whatever is shown on the screen. And for those of you who might be thinking that the price would be too high, just remember that at some of the world's top restaurants individual plates designed for just one dish have been known to come in at more than £1,000 a pop. Plating off a tablet might seem cheap by comparison.

Would you like to eat cheesecake on Mars?

This is what the creators of Project Nourished are offering with their recent attempt to combine virtual reality with food. Developed by Kokiri Labs, in LA, Project Nourished is described as a ' "gastronomical virtual reality experience." This mashup of molecular gastronomy and virtual reality allows users to "experience fine dining without concern for caloric intake and other health-related issues." '[6] Their strapline is 'Wouldn't you like to eat cheesecake on Mars?' Said almost

rhetorically – for how could you resist? You must, at the very least, be
a little curious. And, given what we have seen in previous chapters –
how much impact context, atmosphere and environment have on
dining and drinking – you can be sure that your cheesecake probably
would taste a little different if you were wearing one of these headsets,
assuming, that is, that the experience is suitably immersive, and all
that virtual space dust doesn't blow into your eyes! Looking a little
further into the future, it is interesting to consider how the various
new augmented and virtual reality (AR and VR respectively) technologies are going to enable the diner of tomorrow to eat one food
while simultaneously viewing another.

So what exactly do we all have to look forward to, in this mash-up
of food and virtual reality? Well, here's a hint of what may be to
come, at least if the techno-geeks have their way: 'As for Project
Nourished, here's the deal: You put on your VR headset, [. . .] you
lift your "food detection sensor," which, at this stage of development,
looks like nothing more than a wooden fork with two prongs wrapped
in tinfoil; you eat a hydrocolloid – a flexible substance that is "vis-
cous, emulsifiable, and low caloric" – that has been shaped in a 3-D
printed mold to "add physical characteristics to the 'faux' food."
Then, with the help of a motion sensor, an aromatic diffuser, and a
bone conduction transducer [. . .] you experience a gourmet meal
with no downside in the way of calories, carbs, or allergy-inducing
ingredients.' Don't tell me you are still not convinced!

The more fundamental worry, though, is simply that the idea
of eating cheesecake on Mars doesn't seem like an especially congruent combination, at least not to me. Perhaps we'd be better off
matching the food to the environment that we are immersed in via
the headset, maybe a strawberry-flavoured space cube (i.e., something
that astronauts were given to eat on their space missions)? But then
again, perhaps not, knowing how bad they were rumoured to taste.

Some of the limitations associated with simulating a given environment when restricted to just visual VR are brought home if you
imagine how realistic an experience you would get were you to try
recreating the experience of dining in a plane, for example. You'd
miss the background noise, the lack of air humidity, the lowered

cabin pressure (see 'Airline Food'). You'd also fail to capture the pressure on your knees when the person in front suddenly decides to recline their seat during the meal service. So it really wouldn't be the same kind of experience at all now, would it? Vision is undoubtedly important, but without the other sensory cues, it is unlikely to be all that immersive – at least for those more *extreme* environments that we might want to simulate. I do wonder, though, whether such VR applications might not find a use amongst elderly patients. Perhaps it could be used to take them back to an earlier time, providing visual cues from the past. After all, playing the music of a bygone era has been shown to help enhance consumption.

Augmented reality dining

Augmented reality involves superimposing artificial visual stimuli over the actual scene. So, for instance, the AR system utilized by my colleague Katsuo Okajima and his colleagues in Japan can update the visual appearance of food or drink in real time. Just imagine: you put on the commercially available headset, and at first you can see the sushi that you ordered on the plate in front of you. Then, simply by moving your hand over the plate (abracadabra-style), the fish suddenly changes from tuna to salmon, say. Move your hand over the plate again, and now it is eel. Not only that but you can pick up what looks like eel sushi and even take a bite without destroying the illusion.

So what's the point of it, some of you are no doubt wondering. Well, in some of our preliminary research, we have been able to demonstrate that changing how food looks can result in people saying different things about the taste, as well as the perceived texture, of cake, ketchup and sushi. Here, one could perhaps imagine consumers viewing what looks like highly desirable but unhealthy food while actually eating a healthy alternative. And, looking a few years into the future, I can well imagine how we might all end up craving some virtual sushi, when the seas have been fished to extinction and the real thing is nothing but a dim and distant memory (sorry to be so depressing).

One other intriguing example of the intelligent use of AR headsets while dining comes from researchers who have been looking into whether they can trick you into feeling fuller sooner. They aim to do this simply by making the food that you see through the headset (e.g., a biscuit) look larger than it actually is. However, much though I like the idea of AR and VR dining, my best guess is that we are still some years off seeing these headsets at the dining table, even at the world's most avant-garde restaurants. The limitation here is as much the cost as the fact that it may interfere with the social interaction around a meal.

Have you heard of the 'Sound of the Sea'?

To date, the more immediate uptake of digital technology at the dining table has been linked to the personalized delivery of sound, not vision. Here I am thinking of soundscapes and musical compositions, for instance, that the diner or drinker listens to while enjoying a specific dish or drink. In an earlier chapter, we saw how Heston Blumenthal first became interested in the important, if neglected, role of this sense after we gave him a few sonic chips to nibble on in the Crossmodal Research Laboratory. The chef went away and, together with his talented crew in Bray, started to explore different ways of bringing sound to the table, digitally. The first iteration of 'sonic cutlery' they came up with was trialled 'secretively' for some of the restaurant's regular clientele, but, unbeknownst to the serving staff who were working the pass, a journalist was sitting incognito in the restaurant that day. When he caught wind of the fact that the diners on the *other* table had been served a dish that he himself hadn't had, he immediately summoned the waiting staff over, announced himself and demanded to know what was going on. There was nothing for it but to let him try out the sonic headphones too. The result: a few days later, who should appear staring intently out from the pages of *The Sunday Times* than that same reporter.[7] The all-new eating utensil for the techno-enhanced twenty-first century had just been 'outed'.

Putting the over-ear headphones on tended to mess with the expensive hairdos of a certain section of the clientele, leaving them feeling anything but alluring, and the headphones were unceremoniously

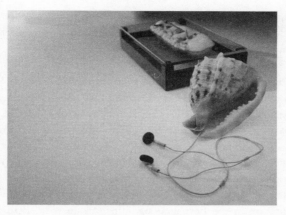

Figure 12.3. The 'Sound of the Sea' seafood dish (for a number of years, the signature dish served on the tasting menu at Heston Blumenthal's Fat Duck restaurant) provides an excellent example of how digital technology can be used to enhance the multisensory dining experience. In research conducted here in Oxford with the chef, we were able to demonstrate that seafood tastes significantly more pleasant (but no more salty) when people listen to waves crashing gently on the beach and seagulls flying overhead than while listening to restaurant cutlery noises or – surprise, surprise – modern jazz!

withdrawn from service almost as soon as they had been placed on the carefully ironed white tablecloths. It was, as they say, time to go back to the drawing board. Roll the clocks forward a couple of years, and what do those who are lucky enough to have booked themselves a seat at The Duck find? Well, for one of the courses, the waiter arrives at the table holding a plate of seafood in one hand: sashimi resting on a 'beach' made of tapioca and breadcrumbs with foam. With his other hand, the waiter passes the diner a seashell from which dangles a pair of MP3 earbuds (see Figure 12.3), and encourages the guest to insert the earbuds before they eat. What the diner hears, assuming that he or she does as told,★ is the sound of the sea: waves

★ According to the team at The Fat Duck, in the eight years or so that this dish has been on the menu, only one diner (a French chef) has not put the earbuds in, saying imperiously that he already knew what the sea sounded like!

crashing on the beach, a few seagulls flying overhead. Some have found the combination of sound and food to be so powerful they have been brought to tears.

Since the 'Sound of the Sea' dish first appeared on the menu in Bray, a number of other chefs (and even the occasional barista)★ have incorporated personalized digital sound into the dishes they serve. For instance, at El Celler de Can Roca, in Girona, part of Spain's *nueva cocina* movement, the culinary team created a dessert that came to the table with an MP3 player and loudspeaker. In this case, diners were encouraged to consume their dessert while listening to a football commentator describing Lionel Messi dodging the Real Madrid defenders and scoring Barcelona's winning goal at the teams' classic confrontation in the Bernabeu stadium back in 2012. Brilliant! It's got both high emotion and storytelling, though I bet it tastes better if you are not a fan of the losing team. (And presumably the dish works better if you actually like football.) Meanwhile, the Michelin-starred chefs at Casamia in Bristol sometimes served a picnic basket with an MP3 player that, when opened in the restaurant, would play the sounds of the English summer.

Surprising spoons

There has also been growing interest in delivering sounds digitally from within the diner's mouth. Chefs, musicians, designers and culinary artists have all become interested in delivering personalized music/ soundscapes to accompany specific tasting experiences. Just take, for example, the limited edition Bompas & Parr baked beans spoons, available for £57. Each spoon had an MP3 player hidden inside. If you bought one, you wouldn't hear anything until you put the spoon into your mouth. Then the sound waves would travel via your teeth and jawbone through to your inner ear. The flavour—music combinations in this case included Cheddar cheese with a rousing bit of Elgar, fiery

★ Top barista Rasmus Helgebostad made a sonically enhanced coffee drink as part of his entry in the 2011 Norwegian barista championships.

chilli with a Latin samba, blues for the BBQ-flavoured beans and Indian sitar music for the curry-flavoured beans! While the diner could hear the music, the person sitting next to them would hear nothing. It remains to be seen just how congruent these musical selections were, and whether they really did enhance the flavour of the food.

Meanwhile, over in the Netherlands, Dutch pianist Karin van der Veen has been offering people the chance to taste digital bonbons, De Muziekbonbon. The idea is simple, really, if a little strange. You put the chocolate, which comes with a wire attached, into your mouth, and as you clench the piezoelectric strip (which vibrates when an electric current is passed through it) embedded in the bonbon between your teeth, you can faintly hear the bone-conducted sound of a piano resonating through your jawbone and carrying on all the way to your inner ear.* A pleasant if most unusual experience, as I am sure you can imagine. While I enjoyed my chance to try one of these multisensory treats, it is certainly not something that I can see going 'mass market' any time soon. I am just not sure that the increased enjoyment is really worth the effort, at least not once you have done it the first time. It is also rather antisocial, in the sense that all conversation is prevented while the 'musical bit' is clenched firmly between your teeth! On the plus side, though, I suppose that you could say that it aids concentration, and hence enhances the experience, nudging the taster towards a more mindful approach to consumption.

Digital flavour delivery

Researchers in Japan have been working to deliver food aromas to match whatever you might see through your AR headset. However, one look at this device (see Figure 12.4) tells you all you need to know about how soon you will be seeing this *beauty* in the modernist restaurant or gadget store. Never!† As is too often the case, the tech-

---

* The thing about bone-conduction is that there tends to be much more emphasis on the lower frequencies of sound.

† Though, it did make an appearance in Japan's entry to Expo '15 in Italy.

Figure 12.4. Hmmm, tasty! Sometimes I worry that human–computer interaction (HCI for short) researchers may spend a little too much of their time thinking about what is possible at the intersection between technology and food and perhaps not enough time considering what is actually likely to be applicable or even desirable out there in the real world. Even the most innovative modernist chef would, I presume, baulk at the idea of having their diners put one of these devices on. And there we were, thinking that over-ear headphones were intrusive!

nology developed to explore the potential connection, or digital interface, with food (or food aromas in this case) fails to consider the aesthetic appeal of their designs. Big mistake!*

The more plausible mainstream delivery of food scents will, I believe, come from plug-ins, like Scentee. This has already been used for one marketing-led intervention in the US, namely, the Oscar Meyer bacon-scented alarm clock app. You simply insert a small

---

* That said, the all-new Nosulus Rift headset can deliver aromas via a very sleek black space-age headset; the only problem is it is exclusively designed to emit one unpleasant smell to go with a *South Park* video game called *The Fractured but Whole* (see http://nosulusrift.ubisoft.com/?lang=en-US#!/introduction).

plug-in device to your mobile and set the time, and you will be woken up by the sound of sizzling bacon and the matching smell! Meanwhile, in Spain, top chef Andoni has been using digital scent delivery to extend the interaction with his diners; those who have booked a visit to Mugaritz can get to experience the actions, aromas and sounds that accompany one of the more multisensory of the dishes on the tasting menu in advance by downloading the appropriate app (see Figure 12.5). On making a circular motion to virtually grind the spices that are visible on the screen of their mobile device, the user not only hears the sound of mortar on pestle but also gets a blast of spicy aroma up their nostrils (via the scent-enabled plug-in). These are exactly the same actions, sounds and aromas that they will subsequently experience on tasting the dish itself while they are sitting in the restaurant. One of the aims is to use multisensory stimulation in order to help build up anticipation in the mind of the diner prior to their arrival at the restaurant. Who knows, the expectant diner may even start to salivate.

However, while such digital smell delivery is certainly practical, the fundamental problem is whether anyone is going to buy the refill. This, in a way, contributed to the demise of DigiScents (a digital smell delivery company started during the internet boom years) a couple of

Figure 12.5. A scent-enabled app helping to build anticipation in the mind of the diner booked into the Mugaritz restaurant in Spain.

decades ago (and at no small expense to investors, it should be said).[8] My best guess is that, while the technology works, there isn't really any consumer desire, or need, for such digital innovation just yet. And without that, these dreams of digital smell are likely to fail, just like those previous attempts.

How would you fancy eating with a vibrating fork?

As we saw back in the 'Touch' chapter, the world of cutlery design is set for a revolution. Part of the change will come from the introduction of new forms, materials and textures for cutlery, but another strand of future development may well revolve around the emergence of digitized, or augmented, eating utensils. For, at least according to the human–computer interaction community, this may also radically transform the way in which we will interact with our food in the years to come. Just imagine a fork that vibrates to let you know that you are gobbling your food a little too fast! No, really! (See Figure 12.6.)

Perhaps the most interesting example of digitally augmented

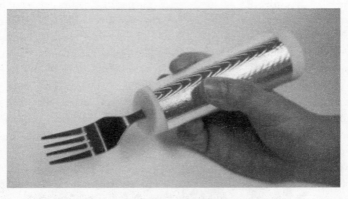

Figure 12.6. One of the ways in which digital technology could make its way on to your dining table in the future. This is an early prototype of the HAPIfork, a Japanese gadget designed to modify our eating behaviour.

cutlery goes by the name of Gravitamine. This utensil cleverly creates the illusion of weight in the hands of its users. Given what we saw in the 'Touch' chapter, I can well imagine how such a digital solution would enhance the diner's meal experience. Though you could be forgiven for wondering whether it wouldn't be more convenient to invest in some genuinely heavy cutlery instead of having to recharge it on a regular basis. One other market where digitally enhanced cutlery may have a promising future is for those patients who find it difficult to control their hand movements – sufferers of Parkinson's disease, for instance, whose tremors can lead to food spillage. In fact, one innovative company has already come out with some anti-shake cutlery to help combat this particular problem.

Electric taste

Researchers can now deliver rudimentary taste sensations simply by electrocuting your tongue in the right way. Any volunteers? Come on now, it isn't as unpleasant as it sounds. Sadly, though, it isn't anything like as pleasurable as many of the press reports would have you believe either! According to some journalists, this all-new approach holds the promise of a never-ending sequence of flavours delivered by your digital device. All you need is a power supply and a stimulator pressed against your tongue. Indeed, the world's press went wild when researchers recently launched a digital lollipop. But hold on a minute, not so fast. The first question to ask before taking all the hype at face value is whether those who are writing about it have actually experienced electric taste for themselves. Too often, this would appear not to be the case! Instead, it seems that much of the time they are relying on second-hand reports by those who are promoting these devices.

I *have* tried some of these devices, and found the experience to be disappointing, to say the least. Now, perhaps I am just unlucky, since electric taste works better on some people's tongues than on others. However, even this approach's most ardent advocates admit that it is easier to get sour and metallic taste sensations than it is

to get salty and umami . . . and sweetness – well, that is a real challenge. Thus the palate of digital tastes that one has to work with, even
in the best-case scenario, is actually pretty limited, and that's for those
in whom electrical stimulation of the tongue works well. I have little
faith that things will improve much, experientially speaking, when
the electrical stimulation device is embedded in the end of a spoon or
in a piece of digital cutlery or glassware.

But more importantly, even if all taste sensations could be rendered
perfectly, you would still be left with a very *thin* dining experience.
For anyone who has evaluated a weak pure tastant in solution will
know only too well how unsatisfactory it is, even in the best of cases.
And as we saw in the chapter on 'Smell', taste constitutes only a very
small part of our multisensory flavour experiences anyway; all those
fruity, floral, meaty, herbal notes that we enjoy while eating and
drinking really come from the nose. In other words, you could never
hope to evoke them by electrocuting anyone's taste buds. You should
be electrocuting their nostrils instead (or as well) – likely to be an
unpleasant, complicated and possibly even painful procedure.

While the original aim of those developing digital taste was to
remove the need to deliver an actual tastant, the approach that is now
being developed involves augmented tasting. Some researchers, for
instance, have been looking at what would happen if your tongue is
zapped while you are looking at a tasty meal, say, or even while eating or drinking something real. It turns out that the hedonic response
to electric taste was indeed changed when people were looking at
gastroporn. Similarly, there is evidence to suggest that people's
response to real food and beverage items can be modified by presenting electrical taste at the same time. So, if diners were given salty
sensations electrically while eating, would this mean that they
wouldn't have to add so much salt? This was the conceit of the 'No
Salt Restaurant', a two-day pop-up in Tokyo. Diners got to eat with
an Electro Fork, capable, apparently, of delivering some electrical seasoning. As a trial run, the restaurant offered a five-course saltless
menu of salad, pork cutlets, fried rice, meatloaf and cake. My guess is
that not many people would have wanted to repeat the experience.

Is this more about marketing than genuine health innovation? One important thing to bear in mind is that the role of salt isn't solely as a taste enhancer. It also plays a key role in determining food texture/structure, and that is something that electric taste simply cannot help with. One further problem with this technology is that our brains seem exquisitely sensitive to how the sensation elicited by different compounds changes over time. This is part of the reason why, for instance, you can distinguish sugar from other artificial sweeteners like aspartame (because the taste sensation ramps up much sooner in one case than the other, and may well linger for longer too). So unless you can get the time-course of sensation associated with electrical taste right, the experience is never going to be as good as 'the real thing'.

Digital technologies changing the food landscape

Go online nowadays and you can find all manner of apps that promise to provide assistance no matter what you want to know about or do to/with your food or drink. Switched-on celebrity chefs and lifestyle bloggers are only too willing to advise you what you should be eating, or else help you to prepare a new dish. This is, after all, big business these days. There are also a growing number of smartphone apps that can interface with, and thus control, various kitchen devices. Just take the 'Bright Grill' as one representative example, an electric barbecue with an app that will alert you as soon as your sausages are cooked perfectly, even if you are away from the grill, so that hopefully you won't end up cremating them, as usual. Don't such inventions make you wonder how we all managed in the good old pre-app days?

In fact, you can find pretty much anything you want at the app store nowadays.* Believe it or not, there is even one called the Egg-Calculator, brought out by ChefSteps. (This is one for all those of you

---

* Sadly, though, the one app that you won't find in the store was developed over in Japan by Kayac Inc. and was based on the results of our sonic chips study (see 'Sound'); Called Evercrisp, it could boost the crunchy, crackly and crispy sounds of the foods you were eating, thus enhancing the eating experience for all.

who are addicted to 'yolk porn' (see 'Sight'); it contains more 'protein in motion' shots than even the most ardent food junkie needs to get off on.) Using this app, there is no longer any excuse for your slow-cooked egg to turn out any other way than exactly how you wanted it. Meanwhile, there are now many price comparison apps that enable the savvy diner to scan their menu and compare how much the same item would cost at other restaurants. On occasion, in big cities like New York, you can find exactly the same bottle of wine being priced at four times what you would pay at another restaurant just down the block. Wouldn't you want to know when you were paying through the nose?

The clever folk at Google have even come up with an app to help restaurant diners split the bill, though the audience for this app is presumably declining somewhat given that more and more of us are dining out alone these days (see 'Social Dining'). And one does have to wonder whether a simple calculator wouldn't do just as well. Interestingly, aware that picking up the hefty tab at a fancy restaurant can put something of a downer on the meal experience of the person paying, and mindful of the fact that the end of an experience tends to be particularly memorable, some high-end restaurants now get their diners to part-pay for the meal in advance, so that the pain at the end isn't quite so bad. That is what I call intelligent design!

Another interesting development involves those sensory apps that allow you to access digital content by scanning the label of anything from a tub of Häagen-Dazs ice cream to a bottle of Krug champagne. For instance, the Concerto app was designed to help the consumer pass the time after taking their ice cream out of the freezer before serving. Customers get their mobile device out and scan the QR code (that black and white patterned square on the lid of special packs), and the next thing you know musicians can be seen and heard 'magically' floating on top of the Häagen-Dazs. Each of the musical selections lasts for about two minutes – i.e., just long enough for the contents to soften slightly, according to the marketing blurb. Once the music draws to a close, the ice cream should be ready to serve.

There are other 'clever' apps that claim to be able to count the calories in a dish by analysing the picture you take of it, and many other new food-related technologies are in development. One project, funded by Philips Research, investigated the feasibility of having people eat from plates that have digital scales embedded in them, in order to calculate the total amount of food that they had eaten. Google's AI, called Im2Calories, is also training itself to count the calories in food photos. It is already accurate to within 20%. But do you really want your technology to keep track of what you eat? What, every gram (or calorie) of it? Furthermore, it remains to be seen quite how accurate these devices really are. After all, the visual system, and the brain that supports it, have been fine-tuned over the course of human history to rapidly assess the energetic content of foods. This is precisely the sort of thing that our brain evolved for, and the evidence suggests that we manage to evaluate nutritive food sources in little more than the blink of the eye. However, even this fine bit of tackle sometimes gets it wrong, or at least our conscious minds do! So why expect the technology to do a better job?

Do robot cooks make good chefs?

But to end, let's return to the question with which we started this chapter: what would you think if you found out that your dinner had been cooked by a robot chef? On the one hand, this should be a wonderful example of precision cooking. That is what we all want, isn't it? Exactly the same taste time after time? And yet, if the food or drink is made by machine, why bother going out to eat in the first place? Why not simply buy it from a packet straight from the production line, as sold in the supermarket? But while the idea of a robot chef, cocktail maker, waiter or even dishwasher might sound completely futuristic, the truth is that the future is already here. For instance, at the Robot Restaurant in Harbin, China, twenty robots, each costing £20–30,000, work the kitchens and restaurant floor. They cook the dumplings and noodles, and wait tables too (see Figure 12.7), though they do need recharging every five hours or so. And KFC recently introduced a robot-serviced restaurant in China. Meanwhile, Royal Caribbean International teamed up with Makr Shakr to

Figure 12.7. Will robot chefs be making our dinner in the future?

create the world's first 'bionic bar' (i.e., a robot cocktail maker) aboard the newest addition to its fleet, the *Quantum of the Seas*. Here's a description of what those who have booked a berth can look forward to while they are out at sea: 'Guests can place drinks orders via tablets and then watch robotic bartenders mix their cocktails. Each robot can produce one drink per minute and up to 1,000 drinks per day, according to Royal Caribbean.'[9]

I was recently approached by the start-up Momentum Machines, who are about to roll out the first robot line/short order chef for mainstream restaurant kitchens. They wanted to know what people would think about food if they knew that it had been prepared by a robot? Would they like it more, or would they perhaps be put off? Do diners really care who makes their food, or are they instead just interested in the taste of the final product? My suspicion is that people will rate food and drink differently (and if I had to guess, less positively) if they are told that it has been created by a robot rather than a real person. As far as I can see, the real sticking point here is likely to be that robots don't have very good tasting abilities.[10] Consequently, such machines will do better when working with packaged (i.e., standardized) ingredients rather than with fresh produce whose quality/ripeness may vary. I also suspect there is something about the predictability of robot-cooking that makes it less appealing than when a real person does all the work.

For better or for worse, then, there really can be little doubting that our future dinners will become increasingly intertwined with digital technologies. This is the case even at home. Indeed, for the home market, Moley Robotics' home-cooking chef should be available early in 2018 for around £50,000.[11] I can just see my wife's eyes lighting up at the prospect.

13. Back to the Futurists

Did you hear about the spat between Paul Pairet and Paco Roncero, two of the world's top modernist chefs, whom we came across in 'The Experiential Meal'? The former accused the latter of stealing his ideas around multisensory experience design. Both chefs currently deliver single-sitting multicourse tasting menus in futuristic dining spaces in which the projections on the walls and table change on a course-by-course basis to complement each of the dishes. Not only that but the music and soundscapes, even the ambient scent and temperature are all designed to match the food. On the surface at least, the chefs' offerings are very similar. They both carefully control the atmosphere in order to deliver a truly multisensory dining *experience*. It is atmospherics taken to the max, facilitated by the latest in technology.

The question remains, though: who deserves the credit? Well, actually, I would say, neither of them do. In this, the final chapter, I want to put the case that as far as modernist cuisine is concerned, the Italian Futurists got there first. They may not have had the knowledge then to bring their dreams to life, at least not in a tasty way, but this is exactly what the science of gastrophysics is increasingly enabling some of the world's top chefs – and eventually you at home – to do. It was the Futurists, after all, who were playing ambient soundscapes to complement the food they served, back in the 1930s; in their case, the sound of croaking frogs to accompany 'Total Rice', a dish of rice and beans, garnished with salami and frogs' legs. Just think about it: Heston Blumenthal only started serving the 'Sound of the Sea' seafood dish, the first multisensory attempt at the Michelin-starred level, back in 2007. And yet something very similar was taking place eighty years earlier, in Turin. No wonder, then, that some people have wanted to argue that the Futurists really were 'Heston's forerunner'.[1]

Similarly, the recent interest of modernist chefs in miscolouring their foods or playing on the assumptions/expectations of the diner – be it Joan and Jordi Roca's completely white dark chocolate sorbet or Blumenthal's own 'Beetroot and Orange Jelly' dish – are again pre-dated by the Futurists. These crazy Italians were deliberately miscolouring various familiar foods in order to discombobulate their dinner guests long before it became trendy for the modernist chef to do so. How would you feel about blue wine,* orange milk or red mineral water? Exactly! They were a dab hand too when it came to brightly coloured cocktails.

The Futurists were interested in touch too, creating what may well be the first painting (entitled *Sudan-Paris*, 1920) designed to be stroked. They had their diners eat without the aid of cutlery, simply burrowing their faces in the plate at their restaurant, the Taverna del Santopalato (the Tavern of the Holy Palate), in Turin. And as we saw in the 'Touch' chapter, it was also in Italy, eighty years ago, that the guests were instructed to stroke their neighbours' pyjamas, made of different materials, while dining.

The interest in fragrance, and in the delivery of food aromas in new and unusual ways, was also something that the Futurists experimented with, waiters spraying atomized perfume directly into the faces of their diners, for example. You find echoes of their exploits in today's modernist food offerings. Chef Homaru Cantu described one of the dishes served at his restaurant Moto, in Chicago, thus: 'This is my favorite part of the meal. I get to pepper-spray our guests.'[2] This reminds me of: 'Aerofood: A signature Futurist dish, with a strong tactile element. Pieces of olive, fennel, and kumquat are eaten with the right hand while the left hand caresses various swatches of sandpaper, velvet, and silk. At the same time, the diner is blasted with a giant fan (preferably an aeroplane propeller) and nimble waiters spray him with the scent of carnation, all to the strains of a Wagner opera.' How multisensory can you get? So, for anyone interested in disruptive multisensory design, a quick look back at the Futurists is perhaps the best place to start.

Just take the rise of 'off the plate' dining and the increasing

---

* Funnily enough, the Spanish company Gïk launched a blue wine in June 2016 (see 'Sight').

theatricalization of service that we came across in 'The Experiential Meal'. By now, you should be able to guess who was there first. As Sophie Brickman notes in *The New Yorker*: 'The banquets and dinners that Marinetti lays out in *The Futurist Cookbook* [. . .] are as much little plays as they are feasts.'[3] Elsewhere, it has been suggested that the Futurists were interested in 'elevating the chef to the rank of sculptor, stage designer, and director of a performative event'.[4] Just take the following account of a dinner hosted in Bologna: 'the "culinary stratosphere" [. . .] was filled with "nutritious noises" of aeroplanes, complementing a mise-en-scène of food sculptures, inventive lighting effects, and amazing outfits for the waiters, designed by Depero.'[5] Remarkably, this event was held on the evening of 12 December 1931. All this, then, leads us to our next question . . .

Futurist cooking: was molecular gastronomy invented in the 1930s?[6]

It is the growing realization of just how many cutting-edge modernist food practices were actually first trialled by Marinetti and his colleagues that has led some to ask whether modernist cuisine really does have its roots back in the 1930s.[7] There are, in fact, a surprising number of similarities between what was happening in northern Italy back then and what one sees in restaurants around the world today. Just take a look at the tenets of the 'Futurist Manifesto' (see below), and then tell me that I am imagining things.

According to the Futurists, then, the perfect meal requires:

1. Originality and harmony in the table setting (crystal, china, decor) extending to the flavours and colours of the foods.
2. Absolute originality in the food.
3. The invention of appetizing food sculptures, whose original harmony of form and colour feeds the eyes and excites the imagination before it tempts the lips.

4. The abolition of the knife and fork for eating food sculptures, which can give prelabial tactile pleasure.

5. The use of the art of perfumes to enhance tasting. Every dish must be preceded by a perfume which will be driven from the table with the help of electric fans.

6. The use of music limited to the intervals between courses so as not to distract the sensitivity of the tongue and palate but to help annul the last taste enjoyed by re-establishing gustatory virginity.

7. The abolition of speech-making and politics at the table.

8. The use in prescribed doses of poetry and music as surprise ingredients to accentuate the flavours of a given dish with their sensual intensity.

9. The rapid presentation, between courses, under the eyes and nostrils of the guests, of some dishes they will eat and others they will not, to increase their curiosity, surprise and imagination.

10. The creation of simultaneous and changing canapés which contain ten, twenty flavours to be tasted in a few seconds. In Futurist cooking these canapés have by analogy the same amplifying function that images have in literature. A given taste of something can sum up an entire area of life, the history of amorous passion or an entire voyage to the Far East.

11. A battery of scientific instruments in the kitchens: ozonizers to give liquids and foods the perfume of ozone; ultra-violet ray lamps (since many foods when irradiated with ultra-violet rays acquire active properties, become more assimilable, preventing rickets in young children, etc.); electrolysers to decompose juices and extracts, etc. in such a way as to obtain from a known product a new product with new properties; colloidal mills to pulverize flours, dried fruits, drugs, etc.; atmospheric and vacuum stills; centrifugal autoclaves; dialysers. The use of these appliances will have to be scientific, avoiding the typical error of cooking foods under steam pressure, which provokes the destruction of active substances (vitamins, etc.) because of the high temperatures. Chemical indicators will take into account the acidity and alkalinity of the sauces and serve to correct possible errors: too little salt, too much vinegar, too much pepper or too much sugar.[8]

We have come across modernist chefs addressing pretty much every one of these issues in the preceding chapters. Indeed, the last point on the list sounds just like molecular gastronomy/modernist cuisine – call it what you will. While the names of the latest must-have kitchen gadgets have undoubtedly changed, the underlying idea is the same – science in the kitchen and the preservation of nutrients/flavour (one of the major selling points of the sous vide technique). I wonder what the Futurists would have made of sous vide, or, for that matter, the anti-griddle, popularized recently by chef Grant Achatz. This new modernist kitchen device can flash-freeze or semi-freeze any food that is placed on its chilled surface.

However, there are also some fundamental differences between what the Futurists were trying to achieve in Italy in the 1930s and what many of today's modernist chefs have in mind. The former were certainly not much interested in making their food taste good; rather, they wanted to provoke, to shock people out of their comfort zone, to stop them revelling in the past (in their ossified cultural and political institutions, as some described it). Today, by contrast, the world's most talented chefs are increasingly realizing that they need to control 'the everything else' in order to deliver the most stimulating, the most memorable and hopefully the most enjoyable dining experiences. The aim nowadays is to prepare the best-tasting food possible and to complement that with the most immersive and engaging multisensory stimulation 'off the plate' as well.

On reading about some of the Futurists' crazier ideas, I am reminded of Albert Einstein's quote: 'If at first the idea is not absurd, then there is no hope for it.'[9] Just take, for example, the incendiary suggestion that pasta should be banned. Marinetti argued that it interfered with critical thinking due to its heaviness in the stomach. He also objected to the fact that it is 'swallowed, not masticated'.[10] Could you imagine a more provocative suggestion in Italy? It is funny, though, to read contemporary descriptions of Futurist dinners, even when presented by sympathetic chefs or journalists; the end result rarely sounds all that palatable. Just take the 'Libyan Aeroplane' dessert: glazed chestnuts marinated in eau de Cologne followed by milk (don't tell me, you're not hungry), served atop a pâté of apples, bananas, dates and sweet peas in the form of an aeroplane – the Futurists, after all, loved their machines (note the steam engine on the wall in Figure 13.1).

There are other differences beyond the Futurists' lack of interest in how the food actually tasted. Marinetti had in mind a future where the calorific requirements of the populace would be met by pills and powders, giving the body 'the calories it needs as quickly as possible'.[11] His idea was that once the basic nutrition was taken care of, this would free up time for 'novel experiences for the mouth and tongue,

Figure 13.1. A Futurist banquet in Tunis, *c.*1931. I must say that the diners (including Filippo Tomaso Marinetti (1876–1944) himself, shown staring intently at the waiter) look much more conventionally attired than all that talk of textured jim-jams would have led one to believe.

as well as for the fingers, nose and ears'.[12] The touch, sound and scent of the Futurists' dishes were actually meant as *substitutions* for the nutritive function of food. Marinetti himself was clear on this point, describing the latter as 'a dish I would not recommend for the hungry'.[13] By contrast, the modernist chef's food is designed to satisfy their guests' hunger, as well as to look good on the plate (i.e., feeding their mind). Though those who remember the heyday of nouvelle cuisine might disagree.

Of course, even Marinetti's ideas about the future of cuisine didn't arise out of nowhere; rather, one needs to look to Apollinaire (another of the Futurists), who hosted a dinner in Paris in September 1912 (see Figure 13.2). He called this new style of cooking *'gastro-astronomisme'*, after the eighteenth-century astronomer Laval. One can already see this focus on feeding the mind: 'In the true style of nouvelle cuisine, these proto-Futurist culinary innovators did not cook in order to fill the stomach but to satisfy the cravings of the mind. Their intention was to create works of art; therefore "it is better not to have any hunger when tasting these new dishes".'[14] This comment clearly prefigures Marinetti's position, and yet it is the latter who remains the undisputed godfather of Futurism.

Fresh violettes without their stems seasoned with lemon juice
Monkfish cooked in eucalyptus
Rare sirloin steak seasoned with tobacco
Barded quail with licorice sauce
Salad seasoned with oil and marc (brandy)
Reblochon cheese seasoned with walnuts and nutmeg
Fruit

Figure 13.2. Menu from a proto-Futurist meal, Guillaume Apollinaire's 'Le Gastro-astronomisme ou la Cuisine Nouvelle' (1912–13). Notice the similarities with nouvelle cuisine (e.g., in the unusual combinations of ingredients). In his book *Feast and Folly*, Allen Weiss explicitly draws parallels with the contemporary cooking practices of a number of top French chefs, including such luminaries as Michel Bras, Pierre Gagnaire and Alain Hacquard.[15]

Want to create your own Futurist party?

Here are my recommendations for those of you who want to go 'off the beaten track' and create your very own Futurist party. Given that modernist chefs have taken so much of their inspiration from the Futurists, there is really no reason why you at home shouldn't do the same. So, my top tips include:

1. Cover your dining table (and if you have enough, the walls too) in aluminium foil. This was an exciting new material at the time the Futurists were doing their thing, both futuristic and technological. (Though it might be difficult to create quite the same sense of wonder in your guests today using foil.)

2. The amuse bouche should probably consist of a dieting pill. Remember, Futurist food is all about feeding the mind, not the body.

3. Buy some atomizers and infuse small amounts of the herbs, spices or fruits that feature in your recipes in oil or water. Encourage your guests to spray before tasting the food, and encourage them to breathe deeply. Perhaps position a fan pointing towards the table and set it on full power. (I am assuming that you don't have a jet propeller to hand.)

4. Use variously textured place mats, or else give everyone a few swatches of materials like sandpaper, velvet and silk. Encourage your guests to rub the full range of materials while eating/drinking to see whether it brings out something different for them in the tasting experience. If any of your guests turn up in a velvet smoking jacket or a silk dress, so much the better! (But that may just be a north Oxford thing.)

5. Between courses, why not put some Wagner on the music system and turn the volume up loud?

6. Create a really fragrant sauce in a pan, then bring it round the table suggestively wafting it under your guests' nostrils before returning the pan to the kitchen untouched.

7. No need to bother with cutlery; simply encourage your guests to eat with their hands and/or burrow their faces into the dishes.

8. Get a selection of food colourings from the supermarket and mischievously add some to each and every one of the drinks you serve.

9. Make use of lashings of those crunchy silver balls used to decorate cakes. Symbolic of the machine age, these tasty cake toppings will be more palatable than the real ball-bearings that the Futurists used to stuff into their 'Chicken Fiat' dish.

10. Make sure you play the appropriate nature sounds to accompany your dishes: the sounds of the sea for seafood, croaking frogs for frogs' legs and mooing cows if you happen to be serving beef. Your guests may never look at a steak in quite the same way again . . .

11. And why not engage in a dash of sonic seasoning? Serve a bittersweet dish, something with dark chocolate or sweetened black coffee, perhaps. Then alternate between tinkling, high-pitched piano music and some low-pitched, brassy music and see whether the taste of the food you have prepared changes.

12. In terms of the food itself, I would go for straightforward nouvelle cuisine rather than full-on Futurist fayre. And if any of your know-it-all guests query why they aren't being offered 'Drum Roll of Colonial Fish', 'Excited Pig' or 'Clotted Blood Soup', or tell you that they were especially looking forward to 'Italian Breasts in the Sunshine', just remind them that Futurist cuisine actually started with Apollinaire's proto-Futurist banquet in Paris, back in 1912!

13. Whatever you do, don't serve PASTA!

You'll be sure to deliver a night to remember.

Looking forward to the future of food

As robot chefs become more popular (see 'Digital Dining'), the question of *who* makes our food, and *how*, will increasingly come to the fore. Similarly, as a growing number of the world's top chefs open multiple restaurants under their own brand name, the question we should all be asking ourselves is what, exactly, are we buying (buying into) when we go to these restaurants? In what sense have the dishes we order really been touched by the master's hand? Of course, we all want consistency, something that is uniformly good; none of us wants to be disappointed by a dish that just isn't up to scratch. And yet, if it were that simple, wouldn't a robot or production line do a better job of producing a consistent outcome? It could even be programmed to imitate the movements and mannerisms of the star chef – but is that really what we want? Certainly, when we discover that our favourite restaurants are buying their food in pre-prepared, we somehow feel disappointed, cheated even. Worryingly, this is something that a growing number of the larger restaurant chains appear to be guilty of these days.

Tim Hayward, writing in the *Financial Times*, talks about the 'cult of inconsistency'. His suggestion is that we should be celebrating variability in the delivery of the foods we eat, not denigrating it (because it shows that there really was a fallible human hand at work in the process of creation). And that, after all, is what we actually all want when we eat out, isn't it? As one salmon smoker Hayward spoke to put it: 'Why would I want to be "consistent"? It's an artisanal product and variation is part of that. It's how people know it's not mass-produced.'[16]

So, as robot chefs and cocktail makers start to appear more frequently, what will we, as diners and drinkers, think? Our views concerning the preparation of food and drink might well change. Here, I am reminded of the Italian biscuit manufacturer who gives their bags of biscuits a natural, handmade feel (despite the fact that all of their products are created on a production line) by having one of the machines cut biscuits that are somewhat different in shape from all the rest. Put a couple of those unusually shaped biscuits in a bag and the consumer is likely to interpret this subtle cue to mean the product is 'made by hand', and perhaps enjoy the experience of eating them more as a result. That, at least, is the idea.

Will we still be so enthusiastic about the open kitchen concept in restaurants once the chef is robotic rather than human? At present, the new technology has novelty value for sure, but for how much longer? And there are a number of other emerging trends that may even presage the disappearance of the restaurant as we know it. Sounds unlikely? Crazy, even? And yet I believe that things may be slowly starting to move in that direction. There's a new presence on the streets of a growing number of cities – the green and black delivery boxes of Deliveroo whizzing around on motorbikes and cycles.* Other companies have gone one stage further; for example, if you live in central London (i.e., Zone 1), Supper will deliver high-end Michelin-starred cuisine direct to your front door. So if, in the years to come, such home delivery services continue to grow at anything like the current pace (and the price comes down, as it always does), then the question on every restaurateur's lips really ought to be: will people continue to go out to eat in restaurants? Why, after all, should they bother, when they can enjoy the same dishes in the comfort of their own homes? Presumably this is exactly the same issue that cropped up when the possibility of watching the latest movies at home, rather than at the cinema, first became a reality.

What, exactly, is lost if you take the restaurant away from the food? Well, as we have seen in earlier chapters, the fact that no crockery, cutlery or napkins are provided by the home delivery services is likely to diminish the at-home experience (assuming that the cutlery

---

* In my local pizza restaurant, Mamma Mia's, in Jericho, north Oxford, for instance, it is becoming increasingly difficult to keep track of all the Deliveroo drivers queuing up to collect their deliveries. In fact, it is hard not to come away from a meal at the restaurant without having formed the impression that the majority of the trade for venues such as this is switching from *on-* to *off-*premises. Of course, you need the bricks-and-mortar establishment in order to distinguish yourself from the straight take-away (and charge a premium as a result).

used in high-end restaurants is likely to be of better quality and heavier than what most of us use at home). So, if you are tempted to try out these services, my tip is to choose your plateware and cutlery carefully. It really can make *all* the difference. Oh, and – as you'll know by now – be sure to get the music right too.

There has also been a recent push towards helping people to make their own food at home. A number of internet-based companies (think Blue Apron, HelloFresh and ChefSteps) now encourage people like you and me to make a chef-prepared recipe by sending you the ingredients in the right proportions and offering a step-by-step guide online on how to prepare the food. Should this trend continue, we might see more home chefs cooking healthier foods. These meals are likely to taste better too, given the Ikea effect we saw in 'The Personalized Meal'.

Another pressure on the traditional restaurant format comes from all those creative chefs who are busily transforming dining into something more akin to a show. Sure, there is still food involved, but it is no longer necessarily the central focus of the experience. Think of it as modernist 'eatertainment', if you will. If you had been to one of chef Jozef Youssef's 'Gastrophysics' dinners in London in 2016, you would have heard the sound of a duck quacking, and then the duck being 'terminated' (imagine the sound of a meat cleaver on a heavy wooden chopping board, complete with the cracking of cartilage and bone), all before the duck course comes out from the kitchen. As the chef puts it: 'If [thinking about where their meal came from] makes the diner uncomfortable then they shouldn't be eating this animal in the first place.' The most important thing here is to keep the diner entertained, but, beyond that, there can be a more serious aim, to nudge the diner towards making healthier, more sustainable food choices.[17]

Already some of the language of dining is changing subtly; 'diners' are gradually being replaced by 'guests'. And what is more, you increasingly find yourself booking a ticket for the show, rather than a table for dinner (see 'The Experiential Meal'). As these trends continue to develop, the restaurant as we know it may soon disappear, or at least evolve into quite a different kind of experience (think of all those coffee shops selling books that morph into bookshops selling coffee).

Big data and food

Looking forward, it is going to be interesting to see how big data and citizen science will change the design of the food experiences that we are all exposed to. We have already started to see linguists mining thousands of online menus to figure out how much we have to pay for each extra letter in a dish name on the menu – around six cents, apparently.[18] And then there are computer scientists out there busily comparing recipes from around the world in order to discover key flavour pairings that are intimately associated with the cuisine of a particular place (or region), ushering in a new area of science, known as 'computational gastronomy'. The latest analysis of Indian recipes, for instance, shows that the chefs there tend to combine ingredients that are not harmonized – exactly the opposite pattern to what is seen in the rest of the world.

What other genuine insights might emerge from this data-mining of large-scale food-relevant databases? Will a whole range of great-tasting but unusual new flavour combinations emerge from the likes of FoodPairing (who run a subscription website allowing chefs, cocktail makers and interested home chefs to figure out which combinations of ingredients share the same flavour compounds) and IBM's Chef Watson? IBM's supercomputer Watson algorithmically analyses a database of thousands of recipes plus a database of flavour compounds found in thousands of ingredients, along with psychological findings about how people perceive different combinations of ingredients. This computer has no hands so merely comes up with unusual combinations that someone else makes: 'IBM is keen to stress that this is not about machines outdoing humans but rather working side by side with them [. . .] Heston Blumenthal better watch out'.[19] Will the diners of tomorrow be increasingly exposed to a whole range of new flavour combinations? The key point to remember here is that it is not about gastrophysicists competing against chefs, nor computers battling it out against humans. Rather, it is a question of how much more convincing a proposition we can make by bringing together the various disciplines. Meanwhile, researchers who surveyed over a million online restaurant reviews over a nine-year period from 2002 to 2011 across every state in the US found that good weather tends to result in our enjoying the experience of dining out that much more than when the weather is not so good.

In the last few years, the Crossmodal Laboratory has been running a number of large-scale citizen-science experiments in museums and also online, to help provide information about the kind of design decisions that diners may appreciate: everything from assessing the orientation of the dish through to the wall colour and the background music. My guess is that the small-scale studies of the effect of the environment on the behaviour of diners (typically involving a few tens or hundreds of diners, as covered in 'The Atmospheric Meal') will soon be eclipsed by big-data studies (with the data perhaps acquired through the signals given off by diners' mobile devices). The number of participants that one can study will suddenly jump to the tens or hundreds of thousands.

This should hopefully allow for much better evidence-based decision-making in the design of food and beverage offerings. As a case in point, over the last year we have collected responses from more than 50,000 people who attended the 'Cravings' exhibition (either in person or online) while it was open to the public in London's Science Museum. The results have helped confirm some of our intuitions regarding how the presentation of food affects what we think about the flavour and how much we expect to like a dish. At the same time, however, the results have started to disprove some kitchen folklore – such as, for example, that odd numbers of items are preferred over even numbers on the plate. We have recently extended this approach to look at the best orientation for a long straight element in a dish – think a seared spring onion, or a whole lobster. It turns out that people like linear elements to ascend from the bottom left to the top right. Another intriguing contrast that the latest research highlights is between the kinds of plating people will be willing to pay most for, and those they deem most creative.

One other example of what is currently being done with big data analysis comes from Rupert Naylor, of Applied Predictive Technologies. He describes what the company he works for does for the restaurant chain as allowing them to: 'run control experiments, much in the same way that they test for the efficacy of new drugs [. . .] We do a control test on restaurants that show similar behaviour as a baseline, then we strip out all the noise – the data that may have affected sales anyway – to get to the truth.' The approach apparently helped Pizza Hut UK to increase average customer spend from £9 to £11. While this might not sound like much, rest assured it soon adds up.[20]

Synaesthetic experience design

Practitioners of multisensory experience design have, over many years, thoroughly explored those connections between the senses that are, in some sense, obvious. Just think of a dish of frogs' legs accompanied by the sound of croaking frogs, or a seafood dish accompanied by the sounds of the sea. Other restaurants, including Eleven Madison Park in New York, Casamia in Bristol and Ultraviolet in Shanghai, have tried various ways to recreate a picnic (or, more likely, the positive emotions that are associated in most people's minds with such events from their past): by using ceramic plates that imitate paper plates; by serving from a picnic basket, obviously; and probably by introducing the sounds, smells (scent of a field) and even associated visuals (think of a countryside scene). It is effective, most definitely, but it can seem a little clichéd, at least in hindsight. Increasingly, though, chefs, culinary artists and experience designers are starting to move more into the world of synaesthetic design, where decisions about the experiences that are delivered to diners (and drinkers) are based on connections between our senses that are not so obvious, or literal. Here, I am thinking of events like 'The Colour Lab', where ambient colours and music were used to alter the taste of a glass of wine. This is synaesthesia-like, in that the connections between the senses are often surprising when first you hear about them (e.g., that sweet is high-pitched, pinky-red and round). And yet this form of design is fundamentally different from synaesthesia (the condition where people see letters, numbers and units of time in colour, say, or where sounds trigger smells). The key difference is that these newly discovered links between the senses tend to be shared between the majority of people. It is these common yet often surprising associations – what are often referred to as 'crossmodal correspondences' – that allow for the creation of multisensory experiences that are both intriguing and truly meaningful. The growing body of gastrophysics research provides a number of insights for chefs and experience designers in this area.

Things begin to get really interesting, though, as soon as one starts playing with the correspondences as they relate to the chemical senses – that's taste, aroma and flavour. This is not to say that adding more senses into the mix necessarily makes things easier as far as multisensory experience design is concerned. Just take Sean Rogg, talking about one of his recent events, part of the Waldorf Project, in which people were 'invited to taste colour'. The visitors, who had been asked to dress monochromatically for the event, got to drink fine wine and watch a host of dancers perform. According to the artist: 'Not only did the soundscapes have to sound like their respective colours, but I was asking the sound designer to pair his soundscapes to wine.'[21] That's a big ask. Nevertheless, despite the inherent challenges associated with working in this area, what is clear is that there has been an explosion of interest in synaesthetic design involving food and drink.

There is no guarantee that different people will necessarily have the same reactions to such experiential events, but that is part of the fun. The rise in synaesthetic design, building on the surprising connections between the senses that we all share, goes hand in hand with the emergence of 'sensploration', the idea that consumers are increasingly curious to explore their own sensory world (or 'sensorium') and the hidden connections that can be found within each and every one of us.[22] And while sensory marketing seemed once to be all about making money, it now feels much more about the delivery of shared (and shareable) multisensory experiences (or at least it should). For the culinary artist, it is as much about a journey of discovery. It is about exposing us all to the unusual, surprising, almost-synaesthetic connections between the senses. In fact, at least according to one recent report: '70 percent of United States-based Millennials now search for experiences that "promote their senses."' There are various explanations for this. One intriguing suggestion is that they are hungry for immersive engaging experiences, and as one commentator put it: 'As consumers grow evermore weary of constant digital bombardment, they seek out more authentic experiences to immerse themselves with a brand.'[23] So, while the experience economy continues to influence so many aspects of the marketplace, and marketing communications, it is time to get ready, I think, for the next 'sensory explosion' (as US marketing professor Aradna Krishna called it in an industry briefing back in 2013).

Have you heard of the Gesamtkunstwerk?

Ultimately, then, we are slowly making our way towards the *Gesamtkunstwerk*, food as a total work of art,[24] an installation, or experience, that engages all of the viewer's senses (if 'viewer' is still the right word?). The term is associated with the German composer Wagner, so no wonder that he was the composer of choice at the Futurists' dinner parties. In fact, it is hard to see how this objective of creating a work of art that stimulates all of the senses could ever be achieved without the involvement of food or drink.

The *Gesamtkunstwerk*, Futurism and various other artistic trends that were very much in vogue a century ago can all be linked, more or less directly, to the rise of the physiological aesthetic, in turn-of-the-century Europe. At that time, artists, including famous painters such as Seurat, were convening with scientists in order to design more pleasing experiences based on the emerging neuroscientific understanding of the mind of the viewer. This interaction between the artists and the scientists undoubtedly led to a phenomenal wave of creativity, albeit one that ultimately fizzled out. The decline was probably due to the fact that it was the *wrong* science at the *wrong* time (brain science has undoubtedly come a long way in the intervening 120 years or so). And, as far as I can see, measuring brainwaves was never going to produce the information that the painters (and other artists) wanted to help them in the design of their compositions.

However, roll the clocks forward and we are now increasingly seeing the merging of culinary artistry with the behavioural or psychological sciences. This is encompassed within the new science of gastrophysics. And when combined with the latest in design and technology, this new collaboration promises to deliver a food future that is utterly unlike anything we have seen before – unlike, even, the wildest dreams and ideas of the Italian Futurists.

Looking forward to a healthy and sustainable food future

It is getting ever harder to think about the future of food without being conscious of the problem of climate change, the challenges associated with sustainability, and the rise of the mega-city. It is hard to say whether the solutions to sourcing our food in the future will come from vertical farming, lab-grown meat, increased entomophagy, or – heaven forbid – Soylent Green (a delicious green wafer advertised as containing high-energy plankton but actually secretly made of human remains). This dystopian prediction about the future of food came from Richard Fleischer's movie of the same name, set in the year 2022.* But I passionately believe that whatever the future holds, we will stand a far greater chance of achieving our goals by exploring the interface between modernist culinary artistry as it meets the latest in technology and design. Ultimately, it is crucial to realize that changing behaviour is not simply a matter of informing people of what is good for them or what is sustainable for the planet. Other

---

* See C. Spence & B. Piqueras-Fiszman *The Perfect Meal: The Multisensory Science of Food and Dining* (Oxford: Wiley-Blackwell, 2014) for a discussion of Soylent Green and the new algal cuisine. The movie was based on Harry Harrison's 1966 novel *Make Room! Make Room!* Spotting a marketing opportunity, a California start-up recently started marketing a food product they call Soylent Green, which supposedly contains all the protein, carbohydrates, lipids and micronutrients that we need. Though what was most noticeable about the early formulations of this product was the extreme flatulence it induced in many of those who tried it.

strategies are needed to nudge people towards healthier and more sustainable food behaviours, approaches that are cognizant of the fact that our perception of food happens mostly in the mind, not in the mouth. My guess is that the notion of 'food hacking' is one that we are all going to become much more familiar with in the future.[25]

And from a more personal perspective, I believe that the future of gastrophysics holds fundamental challenges as well as many opportunities to really make a difference to the way many of us will interact with food in the years to come. I hope that some of the most exciting developments currently seen primarily in the high-end modernist restaurant will be scaled up for the masses; indeed, I am already seeing a huge increase in interest from a number of the world's largest food and beverage companies. As ever more of our food behaviours are taking place or being facilitated online, that will open up a whole new world in terms of the big-data analysis of food trends and behaviours. Chris Young, founder of ChefSteps, predicted that their website would be helping more than a million people to cook better meals at home by the end of 2016. This kind of interaction will generate huge amounts of data that can be used to figure out how best to personalize and enhance our food perception and our behaviour moving forward.

The scientific approach represented by the emerging field of gastrophysics will help quantify what really matters here, separating fact from fiction and intuition. Real progress will be made just as soon as more of us recognize that 'pleasure is not only found in the mouth', as chef Andoni Luis Aduriz put it, but also – mostly, in fact – in the mind. Actually, the chef is worth quoting at length here: 'What it comes down to is: *you don't have to like something for you to enjoy it* or, in other words, pleasure is not only found in you. Predisposition, the ability to concentrate – the impulsive mechanisms of the brain – can completely modify the perception of something that, at first sight, would not even be considered food for humans. In the end, it isn't only about eating; it's also about discovering. We take advantage of the fact that we are always on the borderline between our conservative selves – the part that makes us creatures of habit, finding shelter and security in repetition – and our curious, and daring, selves, which

Figure 13.3. F. T. Marinetti gazing into the future.

seek pleasure in the unknown, in the vertigo we feel when we try something for the first time, in risk and the unpredictability.'[26] And that, I think you will find, ladies and gentlemen, brings us right back to where we started! (See Figure 13.3.)

And to end: how about something healthy?

To close this book, I have summarized some of the key recommendation for anyone who wants to feel more satisfied while consuming less (i.e., to eat more healthily):

1. Eat less – Obvious, I hear you say! But not everyone does it.
2. Hide food – you will be more tempted to snack if you can see the cookies in the jar than if they are in an opaque container. It really is a case of out of sight, out of mind. In fact, anything that you can do to make it more difficult to get your hands on food in the first place is likely to help too. This kind of nudge approach has often (not always, mind) been shown to be effective in helping reduce consumption.

3. Middle-aged and older adults should try drinking lots of water before every meal – half a litre thirty minutes before breakfast, lunch and dinner ought to do the trick. In one study, this led to a reduction in consumption at mealtimes of roughly forty calories. Plus all those extra bathroom breaks will no doubt help increase your physical activity!

4. If you happen to be fond of junk food, why not eat in front of a mirror, or else off a mirrored plate? Research suggests it can help reduce desire for, and consumption of, foods such as chocolate brownies. There is at least one famous actress who would apparently eat naked in front of a mirror. It would be interesting to see whether customers at those naked restaurants that have been popping up recently (see 'Social Dining') also eat less. Try to eat slowly and mindfully. And yes, that does mean turning the TV off!

5. The more food sensations you can muster, the better. Stronger aroma, more texture – it all helps your brain to decide when it has had enough. In one of my favourite studies illustrating this point, people consumed far more calories when drinking apple juice as compared to apple puree, and when eating pureed apple as compared to apples. Exactly the same food in all three cases; all that differs are the textural cues the brain receives about how much it has consumed (and how much chewing is needed). This is much the same reason why you should never use a straw to drink. It eliminates all the orthonasal olfactory cues that are normally such a large part of the enjoyment (see the 'Smell' chapter). Be sure to inhale the aroma of your food frequently; after all, this is where the majority of the pleasure resides. Whatever you do, don't drink iced water *with* your meals. It numbs the taste buds, plain and simple! Some researchers have even gone so far as to suggest that the North American preference for more highly sweetened foods may, in part, be linked to all the iced water they drink at mealtimes.

6. Eat from smaller plates. This technique is especially effective

when people serve themselves. The numbers here are pretty staggering: if you eat from a plate that is twice as big, for example, you are likely to consume as much as 40% more food.

7. Bowl food – eat from a heavy bowl without a rim, and hold the bowl in your hands while you eat; don't leave it sitting on the table. The weight in your hand is likely to trick your brain into thinking that you have consumed more, and you'll feel satiated sooner.

8. Eat from red plateware. In this context, red plates seem to trigger some kind of avoidance motivation.

9. Eat with chopsticks rather than regular cutlery, or try eating with your non-dominant hand, or else with a smaller spoon or fork. Do anything, basically, that makes it more difficult for you to get the food into your mouth. Along much the same lines, artists and designers from thirty-five countries were recently tasked with making tableware that challenged eating norms and so encouraged slower, more mindful eating, at an Amsterdam supper club. Though you'd best watch your teeth if you ever try eating with the nail-filled spoon that one eager contributor created for the event.

10. Oh, and one excellent tip from Yogi Berra: 'You better cut the pizza in four slices because I'm not hungry enough to eat six.'[27]

Notes

Amuse Bouche

1 'Square plates are an "abomination", says *MasterChef* judge William Sitwell', *Daily Telegraph* (*Food & Drink*), 13 May 2014 (http://www.telegraph.co.uk/foodanddrink/10828052/Square-plates-are-an-abomination-says-MasterChef-judge-William-Sitwell.html).
2 I had the 'pleasure' of being on a double bill with Michael on his home turf at a literary festival held in Dartmoor National Park ('The Perfect Meal', Professor Charles Spence & Michael Caines MBE in conversation, Chagford Literary Festival, 15 March 2015).

1. Taste

1 D. P. Hanig, '*Zur Psychophysik des Geschmackssinnes*' ['On the psychophysics of taste'], *Philosophische Studien*, 17 (1901), 576–623; E. G. Boring, *Sensation and Perception in the History of Experimental Psychology* (New York: Appleton, 1942).
2 A. L. Aduriz, *Mugaritz: A Natural Science of Cooking* (New York: Phaidon, 2014), p. 25.
3 S. M. McClure et al., 'Neural correlates of behavioral preference for culturally familiar drinks', *Neuron*, 44 (2004), 379–87.
4 J. Gerard, *The Herball or General Historie of Plants* (1597; Amsterdam: Theatrum Orbis Terrarum, 1974).
5 O. Styles, 'Parker and Robinson in war of words', *Decanter*, 14 April 2004 (http://www.decanter.com/wine-news/parker-and-robinson-in-war-of-words-102172/).
6 C. Sagioglou & T. Greitemeyer, 'Individual differences in bitter taste preferences are associated with antisocial personality traits', *Appetite*, 96 (2016), 299–308; A. Sims, 'How you drink your coffee "could point to

psychopathic tendencies"', *Independent*, 10 October 2015 (https://www.independent.co.uk/news/science/psychopathic-people-are-more-likely-to-prefer-bitter-foods-according-to-new-study-a6688971.html).

2. Smell

1 H. T. Fincks, 'The gastronomic value of odours', *Contemporary Review*, 50 (1886), 680–95.

2 C. Morran, 'PepsiCo thinks its drinks aren't smelly enough, wants to add scent capsules', *Consumerist*, 17 September 2013 (https://consumerist.com/2013/09/17/pepsico-thinks-its-drinks-arent-smelly-enough-wants-to-add-scent-capsules/).

3 F. T. Marinetti, *The Futurist Cookbook*, translated by S. Brill (1932; San Francisco: Bedford Arts, 1989), p. 43.

4 E. Waugh, *Vile Bodies* (London: Chapman & Hall, 1930), pp. 80–81.

5 S. Cuozzo, 'Bland cuisine and atmosphere don't boost Eat's silent dinners', *New York Post*, 23 October 2013 (https:nypost.com/2013/10/23/bland-cuisine-and-atmosphere-dont-boost-eats-silent-dinners).

6 M. G. Ramaekers et al., 'Aroma exposure time and aroma concentration in relation to satiation', *British Journal of Nutrition*, 111 (2014), 554–62.

7 S. Nassauer, 'Using scent as a marketing tool, stores hope it – and shoppers – will linger: How Cinnabon, Lush Cosmetics, Panera Bread regulate smells in stores to get you to spend more', *Wall Street Journal*, 20 May 2014 (http://www.wsj.com/articles/SB10001424052702303468704579573953132979382).

8 A. Robertson, 'Ghost Food: An art exhibit shows how we might eat after global warming. What would you do in a world without cod, chocolate, or peanut butter?', *The Verge*, 18 October 2013 (https://www.theverge.com/2013/10/18/4851966/ghost-food-shows-how-we-might-eat-after-global-warming).

3. Sight

1 Just take C. S. Peirce, writing almost 150 years ago: 'Sight by itself informs us only of colors and forms. No one can pretend that the images of sight are determinate in reference to taste. They are, therefore, so far general that they are neither sweet nor non-sweet, bitter nor non-bitter, having savor nor insipid' ('Some consequences of four incapacities', *Journal of Speculative Psychology*, 2 (1868), 140–57). Or take Helmholtz, who, a decade later, wrote: 'For example, one cannot ask whether sweet is more like red or more like blue' (*The Facts of Perception: Selected Writings of Hermann Helmholtz* (Middletown, CT: Wesleyan University Press, 1878)).

And for the opposite position, see e.g. B. Miller, 'Artist invites public to taste colour in ten-day event with dancers and wine at The Oval', *Culture24*, 3 February 2015 (http://www.culture24.org.uk/art/art516019-artist-invites-public-to-taste-colour-in-ten-day-event%20with-dancers-and-wine-at-the-oval).

2 J. Johnson & F. M. Clydesdale, 'Perceived sweetness and redness in colored sucrose solutions', *Journal of Food Science*, 47 (1982), 747–52.

3 For instance, Lyall Watson writing at the start of the 1970s: 'We have a deep-seated dislike of blue foods. Take a trip through a supermarket and see how many blue ones you can find. They are rare in nature and equally rare in our artificial hunting grounds. No sweet manufacturer ever successfully marketed a blue confection, and no blue soft drink or ice cream appeared on sale for very long' (*The Omnivorous Ape* (New York: Coward, McCann, & Geoghegan, 1971), pp. 66–7).

4 J. Wheatley, 'Putting colour into marketing', *Marketing*, October 1973, 24–9, 67.

5 C. Spence, 'Assessing the influence of shape and sound symbolism on the consumer's response to chocolate', *New Food*, 17 (2) (2014), 59–62.

6 D. Gal, S. C. Wheeler & B. Shiv, *Cross-modal influences on gustatory perception* (2007; unpublished manuscript (https://ssrn.com/abstract=1030197)).

7 G. Van Doorn et al., 'Latté art influences both the expected and rated value of milk-based coffee drinks', *Journal of Sensory Studies*, 30 (2015), 305–15.

8 See C. Michel, C. Velasco & C. Spence, 'Cutlery influences the perceived value of the food served in a realistic dining environment', *Flavour*, 4:27 (2015).

9 B. Crumpacker, *The Sex Life of Food: When Body and Soul Meet to Eat* (New York: Thomas Dunne Books, 2006), p. 143.

10 See http://www.chefjacqueslamerde.com/home/; D. Galarza, 'Revealed: Instagram sensation Jacques La Merde is . . .', *Eater*, 28 January 2016 (https://www.eater.com/2016/1/28/10750642/revealed-instagram-sensation-jacques-la-merde-is).

11 J. Yang, 'The art of food presentation', *Crave*, 2011; cited in C. Spence & B. Piqueras-Fiszman, *The Perfect Meal: The Multisensory Science of Food and Dining* (Oxford: Wiley-Blackwell, 2014), p. 113.

12 A. Cockburn, 'Gastro-porn', *New York Review of Books*, 8 December 1977 (https://www.nybooks.com/articles/1977/12/08/gastro-porn/).

13 Quoted in E. Saner, 'Plate spinning: The smart chef's secret ingredient', *Guardian*, 12 May 2015 (https://www.theguardian.com/lifeandstyle/shortcuts/2015/may/12/plate-spinning-smart-chefs-secret-ingredient-food-on-plate).

14 J. Prynn, 'Age of the Insta-diner: Restaurants drop ban on phones as foodie snaps become the norm', *Evening Standard*, 28 January 2016, p. 27.

15 A. Victor, 'Keep your background blurry, never use a flash and DON'T overuse filters: How to turn your dull food images into Instagram food porn in 12 simple steps', *Daily Mail Online*, 28 April 2015 (http://www.dailymail.co.uk/femail/food/article-3050116/12-tricks-help-beautiful-food-photos-Instagram.html).

16 C. Spence, Q. (J.) Wang & J. Youssef, 'Pairing flavours and the temporal order of tasting', *Flavour*, 6:4 (2017).

17 C. Duboc, 'Munchies presents: Mukbang', *Munchies*, 17 February 2015 (https://munchies.vice.com/videos/munchies-presents-mukbang).

18 The increase in this case was in the region of 10–15%; C. P. Herman, J. M. Ostovich & J. Polivy, 'Effects of attentional focus on subjective hunger ratings', *Appetite*, 33 (2009), 181–93.

19 L. Passamonti et al., 'Personality predicts the brain's response to viewing appetizing foods: The neural basis of a risk factor for overeating', *Journal of Neuroscience*, 29 (2009), 43–51; p. 43.

20 S. Howard, J. Adams & M. White, 'Nutritional content of supermarket ready meals and recipes by television chefs in the United Kingdom. Cross sectional study', *British Medical Journal* (2012), 345:e7607.

21 F. M. Kroese, D. R. Marchiori & D. T. D. de Ridder, 'Nudging healthy food choices: A field experiment at the train station', *Journal of Public Health,* 38 (2016), e133–e137.

22 C. Michel et al., 'A taste of Kandinsky: Assessing the influence of the visual presentation of food on the diner's expectations and experiences', *Flavour,* 3:7 (2014).

23 T. M. Marteau et al., 'Downsizing: Policy options to reduce portion sizes to help tackle obesity', *British Medical Journal* (2015), 351:h5863.

24 C. K. Morewedge, Y. E. Huh & J. Vosgerau, 'Thought for food: Imagined consumption reduces actual consumption', *Science,* 330 (2010), 1530–33.

25 A. Swerdloff, 'Eating the uncanny valley: Inside the virtual reality world of food', *Munchies,* 13 April 2015 (https://munchies.vice.com/en/articles/eating-the-uncanny-valley-inside-the-virtual-reality-world-of-food).

26 Quoted from Max Ehrlich, *The Edict* (London: Severn House, 1972), p. 173.

4. Sound

1 'How microwave meals are now on the menu at dinner parties', *Daily Mail Online,* 22 May 2016 (http://www.dailymail.co.uk/news/article-3603849/Third-guests-claim-not-bothered-served-ready-meal.html).

2 P. Samuelsson, 'Taste of sound – Composing for large scale dinners', keynote presentation given at the Sensibus Festival, Seinäjoki, Finland, 13–14 March 2014; see also C. Spence, 'Music from the kitchen', *Flavour,* 4:25 (2015).

3 See 'The sounds of Massimo Bottura by Yuri Ancarani & Mirco Mecacci' – video, *The New York Times Style Magazine,* 2016 (https://www.nytimes.com/video/t-magazine/100000004708074/massimo-bottura.html?smid=fb-share).

4 M. L. Demattè et al., 'Effects of the sound of the bite on apple perceived crispness and hardness', *Food Quality and Preference*, 38 (2014), 58–64.

5 M. Batali, *The Babbo Cookbook* (New York: Random House, 2002), cited in J. S. Allen, *The Omnivorous Mind: Our Evolving Relationship with Food* (London: Harvard University Press, 2012), p. 8.

6 G. Weiss, 'Why is a soggy potato chip unappetizing?', *Science*, 293 (2001), 1753–4.

7 M. Batali, *The Babbo Cookbook* (New York: Random House, 2002), cited in J. S. Allen, *The Omnivorous Mind: Our Evolving Relationship with Food* (London: Harvard University Press, 2012), p. 8. There is simply too much that people want to label innate – more than can possibly be so.

8 M. Lindstrom, *Brand Sense: How to Build Brands through Touch, Taste, Smell, Sight and Sound* (London: Kogan Page, 2005), p. 12.

9 E. Byron, 'The search for sweet sounds that sell: Household products' clicks and hums are no accident; Light piano music when the dishwasher is done?', *Wall Street Journal*, 23 October 2012 (http://www.wsj. com/articles/SB10001424052970203406404578074671598804116).

10 Indeed, over and above what it does to the diners' experience, background music also plays an important, if often unacknowledged, role in helping to motivate the serving staff. As Colin Lynch, the executive chef of Barbara Lynch Gruppo, which comprises restaurants such as Menton and No. 9 Park, puts it: 'I don't think I've ever worked in a kitchen that didn't have some form of music in it. The whole energy of the kitchen changes. The speed at which people work changes depending what we listen to. During prep, you zone out. You're doing one thing for 45 minutes straight. It helps you keep that rhythm.' See D. First, 'Music to prep by: The tunes they name can lighten or quicken the mood before service', *Boston Globe*, 27 July 2011 (https://www.boston.com/ae/food/restaurants/articles/2011/07/27/food_and_music_are_complements_in_most_kitchens___before_its_time_to_focus_on_service/).

11 G. Keeley, 'Spanish chefs want to take din out of dinner', *The Times*, 4 May 2016, p. 33 (http://www.thetimes.co.uk/article/spanish-chefs-want-to-take-the-din-out-of-dinner-cr3fpcg7p).

12 Quote from hotel manager Edwin Kramer, of London's Edition Hotel, in L. Eriksen, 'Room with a cue', *The Journal*, Autumn 2014, 26–7; p. 27.

5. *Touch*

1 Quoted in G. Berghaus, 'The futurist banquet: Nouvelle Cuisine or performance art?', *New Theatre Quarterly*, 17(1) (2001), 3–17, p. 15.

2 I. Crawford, *Sensual Home: Liberate Your Senses and Change Your Life* (London: Quadrille, 1997).

3 W. Welch, J. Youssef & C. Spence, 'Neuro-cutlery: The next frontier in cutlery design', *Supper Magazine*, 4 (2016), 128–9.

4 Y. Martel, *Life of Pi* (New York: Harcourt, 2001), p. 7.

5 See S. Poole, *You Aren't What You Eat: Fed Up with Gastroculture* (London: Union Books, 2012), pp. 44–5.

6 B. Stuckey, *Taste What You're Missing: The Passionate Eater's Guide to Why Good Food Tastes Good* (London: Free Press, 2012), p. 93.

6. *The Atmospheric Meal*

1 Or as another restaurateur put it: 'Customers seek a dining experience totally different from home, and the atmosphere probably does more to attract them than the food itself.' ('More restaurants sell an exotic atmosphere as vigorously as food', *Wall Street Journal*, 4 August 1965, p. 1; as cited in P. Kotler, 'Atmospherics as a marketing tool', *Journal of Retailing*, 49 (Winter 1974), 48–64; pp. 58–9.)

2 P. Kotler, 'Atmospherics as a marketing tool', *Journal of Retailing*, 49 (Winter 1974), 48–64, p. 48.

3 M. Sheraton, *Eating My Words: An Appetite for Life* (New York: Harper, 2004), p. 172.

4 Quoted in C. Suddath, 'How Chipotle's DJ, Chris Golub, creates his playlists', *Businessweek*, 17 October 2013 (http://www.businessweek.com/articles/2013-10-17/chipotles-music-playlists-created-by-chris-golub-of-studio-orca).

5 C. Buckley, 'Working or playing indoors, New Yorkers face an unabated roar', *The New York Times*, 19 July 2012 (https://www.nytimes.

com/2012/07/20/nyregion/in-new-york-city-indoor-noise-goes-unabated.html?_r=0).

6 Quoted in T. Clynes, 'A restaurant with adjustable acoustics', *Popular Science*, 11 October 2012 (http://www.popsci.com/technology/article/2012-08/restaurant-adjustable-acoustics).

7 A. Shelton, 'A theatre for eating, looking and thinking: The restaurant as symbolic space', *Sociological Spectrum*, 10 (1990), 507–26; p. 522.

8 Quoted in B. Stuckey, *Taste What You're Missing: The Passionate Eater's Guide to Why Good Food Tastes Good* (London: Free Press, 2012), pp. 85–6.

9 A. Shelton, 'A theatre for eating, looking and thinking: The restaurant as symbolic space', *Sociological Spectrum*, 10 (1990), 507–26; p. 525.

10 'Welcome to the experience economy', *Harvard Business Review*, 76(4) (1998), 97–105; p. 104.

11 Quoted in C. Rintoul, 'The next chef revolution', 'Food is the New Internet' blog (https://medium.com/food-is-the-new-internet/the-next-chef-revolution-dfe75f0820d2#.k62loo2e8).

12 J. Bergman, 'Restaurant report: Ultraviolet in Shanghai', *The New York Times*, 10 October 2012 (https://www.nytimes.com/2012/10/07/travel/restaurant-report-ultraviolet-in-shanghai.html).

13 M. Steinberger, *Au Revoir to All That: The Rise and Fall of French Cuisine* (London: Bloomsbury, 2010), p. 78.

14 E. Lampi, 'Hotel and restaurant lighting', *Cornell Hotel and Restaurant Administration Quarterly*, 13 (1973), 58–64, p. 59.

15 David Ashen of D-Ash design, quoted in R. S. Baraban & J. F. Durocher, *Successful Restaurant Design* (Hoboken, NJ: John Wiley & Sons, 2010), p. 236.

7. Social Dining

1 See S. Cockcroft, 'That really IS a Happy Meal! McDonald's staff throw a surprise birthday party for a lonely 93-year-old widower who has gone to McDonald's almost every day since 2013', *Daily Mail Online*, 20 November 2015 (http://www.dailymail.co.uk/news/article-3327184/That-really-Happy-Meal-Lonely-93-year-old-gone-McDonald-s-day-death-wife-thrown-surprise-birthday-party-restaurant.html).

2 'Dinner for one – now that's my kind of date', 14 April 2016 (https://www.theguardian.com/commentisfree/2016/apr/13/dinner-for-one-date-solo-dining-eat?utm_source=esp&utm_medium=Email&utm_campaign=GU+Today+main+NEW+H&utm_term=167009&subid=16021322&CMP=EMCNEWEML6619I2).

3 Quote from H. F. Harlow, 'Social facilitation of feeding in the albino rat', *Journal of Genetic Psychology*, 41 (1932), 211–20, p. 211.

4 See C. Steel, *Hungry City: How Food Shapes Our Lives* (London: Chatto & Windus, 2008), pp. 212–13.

5 K. Davey, 'One in three people go a week without eating a meal with someone else, Oxford University professor finds', *Oxford Mail*, 13 April 2016 (http://www.oxfordmail.co.uk/news/14422266.One_in_three_people_go_a_week_without_eating_a_meal_with_someone_else__Oxford_University_professor_finds/).

6 H. Rumbelow, 'Tired of takeaways? Try supper in a stranger's home with the Airbnb of dining', *The Times* (*Times2*), 19 November 2015, pp. 6–7.

7 Camille Rumani, co-founder of the VizEat site.

8 R. Cornish, 'Din and dinner: Are our restaurants just too noisy?', *Good Food*, 13 August 2013 (http://www.goodfood.com.au/good-food/food-news/din-and-dinner-are-our-restaurants-just-too-noisy-20130805-2r92e.html).

9 The study, conducted by OpenTable, was quoted in A. Victor, 'Table for one, please! Number of solo diners DOUBLES in two years as eating alone is viewed as liberating rather than a lonely experience', *Daily Mail Online*, 13 July 2015 (http://www.dailymail.co.uk/femail/food/article-3156420/OpenTable-study-reveals-number-solo-diners-DOUBLES-two-years.html).

10 W. Smale, 'Your solo dining experiences', *BBC News* (*Business*), 31 July 2014 (https://www.bbc.co.uk/news/business-28542359).

11 Quoted in Nell Frizzell, 'Dinner for one – now that's my kind of date', *Guardian*, 14 April 2016 (https://www.theguardian.com/commentisfree/2016/apr/13/dinner-for-one-date-solo-dining-eat?utm_source=esp&utm_medium=Email&utm_campaign=GU+Today+main+NEW+H&utm_term=167009&subid=16021322&CMP=EMCNEWEML66 19I2).

12 A. S. Levine, 'New York today: Where to eat alone', *The New York Times*, 11 February 2016 (https://www.nytimes.com/2016/02/11/nyregion/new-york-today-where-to-eat-alone.html?_r=0).

13 Van Goor also says that 'eating alone is the most extreme form of feeling disconnected in our culture'. Note that dining at Eenmaal does not seem to be about stopping by for a bite to eat, but *rather* actually making a statement by deliberately booking to eat alone. Both quotes from B. Balfour, 'Tables for one – the rise of solo dining', *BBC News Online*, 24 July 2014 (https://www.bbc.co.uk/news/business-28292651).

14 A. J. N. Rosny, *Le Péruvian à Paris* (1801), quoted in R. L. Spang, *The Invention of the Restaurant* (Cambridge, MA: Harvard University Press, 2000), p. 64.

15 See http://mellajaarsma.com/installations-and-costumes/i-eat-you-eat-me/; S. Smith, *Feast: Radical Hospitality in Contemporary Art* (Chicago: IL: Smart Museum of Art, 2013), pp. 212–19.

16 http://www.marijevogelzang.nl/studio/eating_experiences/Pages/sharing_dinner.html.

17 Quoted in R. Comber et al., 'Not sharing sushi: Exploring social presence and connectedness at the telematic dinner party', in J. H.-J. Choi, M. Foth & G. Hearn (eds.), *Eat, Cook, Grow: Mixing Human–Computer Interactions with Human–Food Interactions* (Cambridge, MA: MIT Press, 2014), pp. 65–79; p. 71.

8. Airline Food

1 K. Kovalchik, '11 things we no longer see on airplanes' (http://mentalfloss.com/article/51270/11-things-we-no-longer-see-airplanes); A. Toffler, *Future Shock* (New York: Random House, 1970), pp. 206–11; C. Spence, 'Tasting in the air: a review', *International Journal of Gastronomy and Food Science*, 9 (2017), pp. 10–15; C. Spence, 'Drinking in the skies', *Class Magazine*, Winter (2017), pp. 66–8.

9. The Meal Remembered

1 L. P. Carbone & S. H. Haeckel, 'Engineering customer experiences', *Marketing Management*, 3(3) (1994), 8–19; p. 8.

2 O. Franklin-Wallis, 'Lizzie Ostrom wants to transform people's lives through their noses', *Wired*, 3 October 2015 (http://www.wired.co.uk/magazine/archive/2015/11/play/lizzie-ostrom-smell); J. Morton, *MedTech Engine*, 6 January 2016 (https://medtechengine.com/article/appetite-stimulation-in-dementia-patients/).

3 J. A. Brillat-Savarin, *Physiologie du goût* [*The Philosopher in the Kitchen/ The Physiology of Taste*] (Brussels: J. P. Meline, 1835); published as *A Handbook of Gastronomy*, trans. A. Lazaure (London: Nimmo & Bain, 1884), p. 14.

10. The Personalized Meal

1 J. A. Heidemann, 'You've been Googled — bon appetit!', *Chicago Business*, 29 June 2013 (https://www.chicagobusiness.com/article/20130629/ISSUE03/306299997/youve-been-googled-bon-appetit); S. Craig, 'What restaurants know (about you)', *The New York Times*, 4 September 2012 (https://www.nytimes.com/2012/09/05/dining/what-restaurants-know-about-you.html?pagewanted=all&_r=0).

2 Quotes from A. Sytsma, 'Hardcore coddling: How Eleven Madison Park modernized elite, old-school service', *Grub Street*, 9 April 2014 (http://www.grubstreet.com/2014/04/eleven-madison-park-foh-staff-detailed-look.html?mid=huffpost_lifestyle).

3 'Lunchtime poll – investigating patrons', *CNN*, 10 August 2010 (https://cnneatocracy.wordpress.com/2010/10/28/lunchtime-poll-investigating-patrons/); quote from S. Craig, 'What restaurants know (about you)', *The New York Times*, 4 September 2012 (https://www.nytimes.com/2012/09/05/dining/what-restaurants-know-about-you.html?pagewanted=all&_r=0).

4 S. Miles, '6 tools restaurants can use for better guest intelligence', *Streetfight*, 22 July 2013 (http://streetfightmag.com/2013/07/22/6-tools-restaurants-can-use-for-better-guest-intelligence/).

5 Based on an analysis of a huge number of restaurant menus posted online, linguist Dan Jurafsky notes that 'expensive restaurants ($$$$) have half as many dishes as cheap ($) restaurants' (D. Jurafsky, *The Language of Food: A Linguist Reads the Menu* (New York: Norton, 2014), p. 12).

6 'Menus without choice blaspheme against the doctrine of dining', *FT Weekend Magazine*, 23 January 2016, p. 12.

7 Ogilvy's talk is summarized at https://www.warc.com/Content/News/N34910_Behavioural_economics_is_effective___.content?PUB=Warc%20News&CID=N34910&ID=00be1349-4c3d-4b81-81e3-31f01402d325&q=sutherland&qr.

8 Not just any old marketing executive either: it was Ernst Dichter, one of Louis Cheskin's long-term collaborators. Both were emigrés who fled the chaos and persecution in central Europe in the middle of the last century. See: L. R. Samuel, *Freud on Madison Avenue: Motivation Research and Subliminal Advertising in America* (Oxford: University of Pennsylvania Press, 2010).

9 F. T. Marinetti, 'Nourishment by Radio', in F. T. Marinetti, *The Futurist Cookbook*, translated by S. Brill (1932; San Francisco: Bedford Arts, 1989), p. 67.

11. The Experiential Meal

1 Quotes from A. L. Aduriz, *Mugaritz: A Natural Science of Cooking* (New York: Phaidon, 2014), p. 18; J. Simpson & J. Mattson, 'TV chef's grubby steakhouse mixed raw and cooked meat', *The Times*, 26 May 2014, p. 18 (http://www.thetimes.co.uk/tto/news/uk/article4100051.ece).

2 Quotes from L. Collins, 'Who's to judge? How the World's 50 Best Restaurants are chosen', *The New Yorker* (*Annals of Gastronomy*), 2 November 2015 (https://www.newyorker.com/magazine/2015/11/02/whos-to-judge).

3 J. Kinsman, 'Give us a butcher's . . . for diners, seeing is believing', *Independent on Sunday*, 7 June 2015, p. 59.

4 S. K. A. Robson, 'Turning the tables: The psychology of design for high-volume restaurants', *Cornell Hotel and Restaurant Administration Quarterly*, 40(3) (1999), 56–63, p. 60.

5 Quoted in G. Ulla, 'Grant Achatz plans to "overhaul the experience" at Alinea', Eater.com, 23 November 2011 (https://eater.com/archives/2011/11/23/grant-achatz-planning-major-changes-at-alinea.php#more).

6 Quoted in L. Collins, 'Who's to judge? How the World's 50 Best Restaurants are chosen', *The New Yorker* (*Annals of Gastronomy*), 2 November 2015 (https://www.newyorker.com/magazine/2015/11/02/whos-to-judge).

7 J. Bergman, 'Restaurant report: Ultraviolet in Shanghai', *The New York Times*, 3 October 2012 (https://www.nytimes.com/2012/10/07/travel/restaurant-report-ultraviolet-in-shanghai.html).

8 Quoted in M. Joe, 'Dishing it out: Chefs are offering diners a multisensory experience', *South China Morning Post*, 10 January 2014 (https://www.scmp.com/magazines/style/article/1393915/dishing-it-out-chefs-are-offering-diners-multisensory-experience).

9 S. Pigott, 'Appetite for invention', *Robb Report*, May 2015, 98–101, p. 99.

10 Roncero boasts that his is 'the first gastronomic show in the world' (quoted in B. Palling, 'Fork it over: Are the world's priciest restaurants worth the expense?', *Newsweek*, 4 December 2015 (http://www.pressreader.com/usa/newsweek/20151204/282089160685916)). See also A. Jakubik, 'The workshop of Paco Roncero', *Trendland: Fashion Blog & Trend Magazine*, 23 July 2012 (https://trendland.com/the-workshop-of-paco-roncero/).

11 This was one of Grant Achatz's ideas when he talked about overhauling the experience at Alinea in 2011 (G. Ulla, 'Grant Achatz plans to "overhaul the experience" at Alinea', Eater.com, 23 November 2011 (https://eater.com/archives/2011/11/23/grant-achatz-planning-major-changes-at-alinea.php#more)).

12 J. Gordinier, 'A restaurant of many stars raises the ante', *The New York Times*, 27 July 2012 (https://www.nytimes.com/2012/07/28/dining/eleven-madison-park-is-changing-things-up.html).

13 Quotes from J. Rayner, 'Blue sky thinking', *Observer Food Monthly*, 23 August 2015, pp. 18–22, pp. 21–22.

14 J. Gerard, 'Heston Blumenthal: My new Alice in Wonderland menu', *Daily Telegraph*, 1 July 2009 (http://www.telegraph.co.uk/foodanddrink/

restaurants/5700481/Heston-Blumenthal-my-new-Alice-in-Wonderland-menu.html).

15 K. Sekules, 'Food for thought. Copenhagen's coolest dinner the-atre', *The New York Times*, 19 January 2010 (http://tmagazine.blogs.nytimes.com/2010/01/19/food-for-thought-copenhagens-coolest-dinner-theater/).

16 A. Soloski, 'Sleep No More: From avant garde theatre to commercial blockbuster', *Guardian*, 31 March 2015 (https://www.theguardian.com/stage/2015/mar/31/sleep-no-more-avant-garde-theatre-new-york). Felix Barrett is quoted in the article thus: ' "What we're doing with the bar and the restaurant are experiments, research," he said. "How do you tell a story through food? How do you have a three-course meal that has a narrative?", See also 'Sleep No More adds high-end restaurant to its New York roster', *Guardian*, 26 November 2013.

17 S. Mountfort, 'Like Heston meets Crystal Maze', *Metro*, 9 December 2015, p. 49.

18 P. McCouat, 'The Futurists declare war on pasta', *Journal of Art in Society*, 2014 (http://www.artinsociety.com/the-futurists-declare-war-on-pasta.html).

19 C. A. Jones (ed.), *Sensorium: Embodied Experience, Technology, and Contemporary Art* (Cambridge, MA: MIT Press, 2006), p. 19.

20 J. Klein, 'Feeding the body: The work of Barbara Smith', *PAJ: A Journal of Performance and Art*, 21(1) (1999), 24–35, p. 25.

21 J. Finkelstein, *Dining Out: A Sociology of Manners* (New York: New York University Press, 1989), p. 68.

22 Quoted in J. Gordinier, 'A restaurant of many stars raises the ante', *The New York Times*, 27 July 2012 (https://www.nytimes.com/2012/07/28/dining/eleven-madison-park-is-changing-things-up.html).

12. *Digital Dining*

1 Others have expressed similar reservations: 'Media technologies theor-ist Henry Jenkins (2006) would be sceptical of the notion that a new technology such as the PFP [Personal Food Printer] would displace current technologies, collapsing all kitchen appliances into a single

all-mighty black box. Jenkins refers to this as the black-box fallacy' (quoted in G. Hearn & D. L. Wright, 'Food futures: Three provocations to challenge HCI', in J. H.-J. Choi, M. Foth & G. Hearn (eds.), *Eat, Cook, Grow: Mixing Human–Computer Interactions with Human– Food Interactions* (Cambridge, MA: MIT Press, 2014), pp. 265–78, pp. 273–4).

2 D. Meyer, *Setting the Table: Lessons and Inspirations from One of the World's Leading Entrepreneurs* (London: Marshall Cavendish International, 2010), p. 93.

3 B. London, 'World's first sensory restaurant for BABIES complete with digital menus and interactive menus opens doors', *Daily Mail Online*, 5 June 2014 (http://www.dailymail.co.uk/femail/article-2649367/Worlds-sensory-restaurant-BABIES-complete-digital-menus-interactive-menus-opens-doors.html).

4 C. Spence, 'Multisensory marketing' presentation, Zeitgeist Curator, Berlin, 30 August 2012.

5 Quoted in S. Pigott, 'Appetite for invention', *Robb Report*, May 2015, 98–101.

6 A. Swerdloff, 'Eating the uncanny valley: Inside the virtual reality world of food', *Munchies*, 13 April 2015 (https://munchies.vice.com/en/articles/eating-the-uncanny-valley-inside-the-virtual-reality-world-of-food).

7 B. Dowell, 'Listen, this food is music to your ears', *The Sunday Times*, 29 August 2004 (http://www.thesundaytimes.co.uk/sto/news/uk_news/article236417.ece).

8 C. Platt, 'You've got smell', *Wired*, 1 November 1999 (https://www.wired.com/1999/11/digiscent/); A. Dusi, 'What does $20 million burning smell like? Just ask DigiScents!', *StartupOver*, 19 January 2014 (http://www.startupover.com/en/20-million-burning-smell-like-just-ask-digiscents/).

9 S. Curtis, 'Robotic bartender serves up drinks on world's first "smart ship": Royal Caribbean's *Quantum of the Seas* is the most technologically advanced cruise ship in the world', *Daily Telegraph*, 1 November 2014 (http://www.telegraph.co.uk/technology/news/11198509/Robotic-bartender-serves-up-drinks-on-worlds-first-smart-ship.html).

10 T. Fuller, 'You call this Thai food? The robotic taster will be the judge', *The New York Times*, 29 September 2014, A1 (https://www. nytimes.com/2014/09/29/world/asia/bad-thai-food-enter-a-robot-taster. html?_r=0).

11 R. Burn-Callender, 'The robot chef coming to a kitchen near you', *Daily Telegraph*, 6 October 2015 (http://www.telegraph.co.uk/finance/business club/11912085/The-robot-chef-coming-to-a-kitchen-near-you.html).

13. Back to the Futurists

1 See B. McFarlane and T. Sandham, 'Back to the Futurism', *The House of Peroni*, 2016 (https://thehouseofperoni.com/ie-en/lifestyle/back-futurism/); C. Spence, 'Futurist cocktails', *The Cocktail Lovers*, 23 (2017), pp. 22–5.

2 The dish in question was a chilled citrus soup, finished at the table by the waiter spraying a little togarashi mist over the bowl; see P. Vettel, *Good Eating's Fine Dining in Chicago* (Chicago: Agate Digital, 2013).

3 S. Brickman, 'The food of the future', *The New Yorker*, 1 September 2014 (https://www.newyorker.com/culture/culture-desk/food-future).

4 G. Berghaus, 'The futurist banquet: Nouvelle Cuisine or performance art?', *New Theatre Quarterly*, 17(1) (2001), 3–17, p. 15.

5 G. Berghaus, 'The futurist banquet: Nouvelle Cuisine or performance art?', *New Theatre Quarterly*, 17(1) (2001), 3–17, p. 10.

6 This is the title of a recent article; see 'Futurist cooking: Was molecular gastronomy invented in the 1930s?', *The Staff Canteen*, 25 April 2014 (https://www.thestaffcanteen.com/Editorials-and-Advertorials/futurist-cooking-was-molecular-gastronomy-invented-in-the-1930s).

7 Marinetti published his infamous 'Manifesto of Futurist Cooking' in the *Gazzetta del Popolo* in Turin on 28 December 1930 (reprinted in F. T. Marinetti, *The Futurist Cookbook*, translated by S. Brill (1932; San Francisco: Bedford Arts, 1989), pp. 33–40).

8 See S. Smith (ed.), *Feast: Radical Hospitality in Contemporary Art* (Chicago: IL: Smart Museum of Art, 2013), p. 35 for this reproduction.

9 D. MacHale, *Wisdom* (London: Prion, 2002); from https://www-history.mcs.st-andrews.ac.uk/Quotations/Einstein.html.

10 D. Darrah, 'Futurist's idea on food finds Italy contrary', *Chicago Daily Tribune*, 11 December 1931; H. B. Higgins, 'Schlurrrp!: The case for and against spaghetti', in S. Smith (ed.), *Feast: Radical Hospitality in Contemporary Art* (Chicago: IL: Smart Museum of Art, 2013), pp. 40–47; P. McCouat, 'The Futurists declare war on pasta', *Journal of Art in Society*, 2014 (http://www.artinsociety.com/the-futurists-declare-war-on-pasta.html); R. Golan, 'Ingestion/Anti-pasta', *Cabinet*, 10 (2003), 1–5.

11 F. T. Marinetti, *The Futurist Cookbook*, translated by S. Brill (1932; San Francisco: Bedford Arts, 1989), p. 65.

12 H. B. Higgins, 'Schlurrrp!: The case for and against spaghetti', in S. Smith (ed.), *Feast: Radical Hospitality in Contemporary Art* (Chicago: IL: Smart Museum of Art, 2013), pp. 40–47, p. 43.

13 F. T. Marinetti, *The Futurist Cookbook*, translated by S. Brill (1932; San Francisco: Bedford Arts, 1989), p. 84.

14 G. Berghaus, 'The futurist banquet: Nouvelle Cuisine or performance art?', *New Theatre Quarterly*, 17(1) (2001), 3–17, pp. 8–9.

15 In *Le Poète assassiné* (1916; Paris: Gallimard, 1992), pp. 258–9, reprinted and translated in A. S. Weiss, *Feast and Folly: Cuisine, Intoxication and the Poetics of the Sublime* (Albany, NY: State University of New York Press, 2002), pp. 114–15, pp. 145–6.

16 T. Hayward, 'The cult of inconsistency', *FT Weekend Magazine*, 10 October 2014 (https://www.ft.com/content/41cb3e4c-4e66-11e4-bfda-00144feab7de).

17 C. Spence & J. Youssef, 'Constructing flavour perception: From destruction to creation and back again', *Flavour*, 5:3 (2016). The meal in question was coordinated by Kitchen Theory (https://gastrophysics.co.uk).

18 D. Jurafsky, *The Language of Food: A Linguist Reads the Menu* (New York: Norton, 2014).

19 Quote from J. Wakefield, 'What would a computer cook for dinner?', *BBC News Online*, 7 March 2014 (https://www.bbc.co.uk/news/technology-26352743).

20 Quoted in M. Wall, 'From pizzas to cocktails the data crunching way', *BBC News*, 18 August 2015 (https://www.bbc.co.uk/news/business-33892409).

21 B. Miller, 'Artist invites public to taste colour in ten-day event with dancers and wine at The Oval', *Culture 24*, 3 February 2015 (http://www.culture24.org.uk/art/art516019-artist-invites-public-to-taste-colour-in-ten-day-event%20with-dancers-and-wine-at-the-oval).

22 D. Arroche, 'Never heard of Sensploration? Time to study up on epicure's biggest luxury trend', *LuxeEpicure*, 22 December 2015 (https://www.justluxe.com/lifestyle/dining/feature-1962122.php).

23 Quotes from Y. Arrigo, 'Welcome to the booming experience economy', *Raconteur (Future of Events & Hospitality)*, 362 (2016), 2–3.

24 'Through reinventing the overall experience as *Gesamtkunstwerk*, high-end chefs truly claim their place, as Carême articulated, in the pantheon of great artists' (quoted in J. Abrams, 'Mise en plate: The scenographic imagination and the contemporary restaurant', *Performance Research: A Journal of the Performing Arts*, 18(3) (2013), 7–14, p. 14).

25 J. Wapner, 'The flavor factory: Hijacking our senses to tailor tastes', *New Scientist*, 3 February 2016 (https://www.newscientist.com/article/2075674-the-flavour-factory-hijacking-our-senses-to-tailor-tastes/).

26 A. L. Aduriz, *Mugaritz: A Natural Science of Cooking* (New York: Phaidon, 2014), pp. 42–3.

27 Retrieved June 2016, from https://www.brainyquote.com/quotes/authors/y/yogi_berra.html.

Illustration Credits

Figure 0.1: Reproduced with kind permission of Restaurant Denis Martin

Figure 0.2: Courtesy of the author

Figure 0.3: Courtesy of the Science Museum, London

Figure 0.4: © Andy T. Woods, Charles Michel & Charles Spence, 2016

Figure 1.1: © National Academy of Sciences of the USA, 2008

Figure 1.2: © Oxford University Press

Figure 2.1: 'Jelly of Quail' © Ashley Palmer-Watts, reproduced with kind permission of Lotus PR and The Fat Duck

Figure 2.2: The Viora lid reproduced with kind permission of Barry Goffe; Crown's 360End™can reproduced with kind permission of Cormac Neeson

Figure 2.3: © A. Dagli Orti/DEA/Getty Images

Figure 2.4: © PARS International Corp, 2017

Figure 2.5: Courtesy of the author

Figure 3.1: Courtesy of the author

Figure 3.3: © Luesma & Vega SL

Figure 3.4: Foodography campaign created by BBR Saatchi & Saatchi on behalf of the Carmel Winery

Figure 3.5: C. Michel et al., 'Rotating plates: Online study demonstrates the importance of orientation in the plating of food', *Food Quality and Preference*, 44 (2015), 194–202

Figure 3.6: © Roger Stowell/Getty Images

Figure 3.7: Reproduced with kind permission of KEEMI

Figure 3.8: C. Michel et al., 'Rotating plates: Online study demonstrates the importance of orientation in the plating of food', *Food Quality and Preference*, 44 (2015), 194–202

Figure 4.1: Reproduced with kind permission of Massimiliano Zampini

Figure 4.2: © HOANG DINH NAM/AFP/Getty Images

Figure 4.3: © Frito-Lay North America, Inc., 2017

Figure 4.4: Reproduced by kind permission of Naoya Koizumi

Figure 4.5: The Krug Shell, reproduced by kind permission of Krug Maison de Champagne

Figure 5.1: 'Tableware as Sensorial Stimuli, Rear Bump Spoon for Enhancing Colour & Tactility', Ceramic, 2012, courtesy of Jinhyun Jeon

Figure 5.2: Mulberry Textured Sensory Spoons, courtesy of Studio William

Figure 5.3: Courtesy of the author

Figure 5.4: Meret Oppenheim, *Object* (1936) © Artists Rights Society (ARS), New York / Pro Litteris, Zurich, 2017; rabbit spoon reproduced with kind permission of Charles Michel

Figure 5.5: 'Counting Sheep' © John Carey, reproduced with kind permission of Lotus PR and The Fat Duck

Figure 5.6: Reproduced with kind permission of Marcel Buerkle

Figure 6.2: © Space Copenhagen, 2012

Figure 6.3: © Cornell University, 1999

Figure 7.1: *Lonely* © Jon Krause

Figure 7.2: Mella Jaarsma, *I Eat You Eat Me* (2000). Performed at 'Feast: Radical Hospitality in Contemporary Art', Smart Museum, Chicago, 2012. Photography: Smart Museum. Courtesy of the artist

Figure 7.3: Marije Vogelzang, *Sharing Dinner* (Tokyo, 2008). Photography: Kenji Masunaga. Reproduced by kind permission of the artist

Figure 8.1: © The SAS Museum, Oslo Airport, Norway

Figure 8.2: © The SAS Museum, Oslo Airport, Norway

Figure 9.1: Menu map copyright © Dave McKean, reproduced with kind permission of Lotus PR and The Fat Duck

Figure 10.1: © The Coca-Cola Company, 2017

Figure 10.2: © *Chicago Tribune*, 2012. All rights reserved. Distributed by Tribune Content Agency. Photography: Scott Strazzante

Figure 10.3: 'Sweet Shop' © John Carey. Reproduced with kind permission of Lotus PR and The Fat Duck

Figure 11.1 © Alex Lentati

Figure 11.2: Underwater restaurant © Crown Company PVT Ltd trading as Conrad Maldives Rangali Island, 2013; Dinner in the Sky, Toronto © Dinner in the Sky

Figure 11.3: © David Ramos/Getty Images

Figure 11.4: © Liz Ligon

Figure 11.5: Barbara Smith, *Ritual Meal* (1969). Excerpt from 16mm film by William Ransom and Smith of a performance event in Brentwood, California. Lent by the artist

Figure 12.1: © Food Ink, 2016

Figure 12.2: © Charles Spence and Piqueras-Fiszman; licensee BioMed Central Ltd, 2013

Figure 12.3: 'Sound of the Sea' © Ashley Palmer-Watts, reproduced with kind permission of Lotus PR and The Fat Duck

Figure 12.4: © Association for Computing Machinery, Inc., 2012

Figure 12.5: © Intellect Limited

Figure 12.6: © Association for Computing Machinery, Inc., 2011

Figure 12.7: © REUTERS/Sheng Li

Figure 13.1: 'The Futurist Table', *c.*1931 (Filippo Tommaso Marinetti Papers, Beinecke Rare Book & Manuscript Library, Yale University) © DACS Author photograph © akg-images/MPortfolio/Electra

Figure 13.3: © akg-images/MPortfolio/Electra

While every effort has been made to trace copyright holders, the publishers will be happy to correct any errors of omission or commission at the earliest opportunity

Index

He just wanted a decent book to read ...

Not too much to ask, is it? It was in 1935 when Allen Lane, Managing Director of Bodley Head Publishers, stood on a platform at Exeter railway station looking for something good to read on his journey back to London. His choice was limited to popular magazines and poor-quality paperbacks – the same choice faced every day by the vast majority of readers, few of whom could afford hardbacks. Lane's disappointment and subsequent anger at the range of books generally available led him to found a company – and change the world.

'We believed in the existence in this country of a vast reading public for intelligent books at a low price, and staked everything on it'
Sir Allen Lane, 1902–1970, founder of Penguin Books

The quality paperback had arrived – and not just in bookshops. Lane was adamant that his Penguins should appear in chain stores and tobacconists, and should cost no more than a packet of cigarettes.

Reading habits (and cigarette prices) have changed since 1935, but Penguin still believes in publishing the best books for everybody to enjoy. We still believe that good design costs no more than bad design, and we still believe that quality books published passionately and responsibly make the world a better place.

So wherever you see the little bird – whether it's on a piece of prize-winning literary fiction or a celebrity autobiography, political tour de force or historical masterpiece, a serial-killer thriller, reference book, world classic or a piece of pure escapism – you can bet that it represents the very best that the genre has to offer.

Whatever you like to read – trust Penguin.